CONTENTS

Preface

The study of higher mathematics requires the ability to understand and construct proofs. As students pursue the study of calculus, their focus is primarily on computational methods Historically, there has been a gap between calculus studies and courses in higher mathematics such as abstract algebra and real analysis. To bridge this gap and to better prepare students for more abstract mathematics, college and university curriculums have introduced courses that are called transition courses.

In these transition courses, students are introduced to proof techniques and methods of proof. Basic concepts of number systems, sets, relations and functions are usually included in the course material. These concepts provide a context in which proof techniques can be applied. The basic principles of logic are also developed since these are fundamental for understanding proof methods.

This text is designed to provide a comprehensive foundation for the study of advanced mathematics. It is well suited for use in a transition course. Concepts in elementary number theory, set theory, relations and functions as well as limits and continuity are developed. The first chapter presents the properties of the integers, and the rational and real number systems so that students can begin writing proofs in these subject areas early on.

For students, who are planning to teach mathematics at the secondary school level, the rigorous treatment of topics in basic number theory and the real number system provide an excellent background.

To address the difficulties that students experience when working with quantifiers, there is a complete chapter on quantifiers and proofs using these.

Numerous examples of complete proofs are presented throughout the text. When a particular method of proof (such as proof by contradiction) is introduced, this method is then illustrated by a detailed example. The exercise sets are comprehensive and solutions to selected exercises are given in detail so that students can check their understanding.

A major goal is to help students learn how to construct proofs that are not only mathematically correct but are written in a clear and understandable fashion.

Acknowledgements

This book was developed from my experiences in teaching a foundations course for several years at Florida A & M University. Preliminary versions of the book were also used by other instructors, who taught the same type of course. In particular, I wish to acknowledge the support of Professors Charles Foster and Andrew Jones and thank them for using an earlier version of this book as the required textbook in their course sections.

CHAPTER 1 PROPERTIES OF NUMBER SYSTEMS AND INTRODUCTION TO PROOF

The primary number system used in algebra and calculus is the real number system. We usually use the symbol $\mathbb{R}$ to represent the set of all real numbers. The real numbers consist of the rational numbers and the irrational numbers. The rational numbers are those real numbers that can be written in the form $\frac{m}{n}$, where m,n are integers with $n \neq 0$. We use the symbol $\mathbb{Q}$ to represent the set of all rational numbers. The set of integers consist of zero, the positive whole numbers, and the negative whole numbers. We use the symbol $\mathbb{Z}$ to represent the set of all integers. If n is an integer, we can write $n = \frac{n}{1}$ and hence each integer is a rational number.

Properties of the Integers *

The set $\mathbb{Z}$ has the arithmetic operations of addition $+$ and multiplication $\cdot$ defined on its elements. We will assume that the following properties are true for addition and multiplication of integers m,n,p.

Closure Properties & Uniqueness

$m+n$ is a unique integer and $m \cdot n$ (written mn) is a unique integer

Commutative Properties

$m+n = n+m$, $mn = nm$

Associative Properties

$(m+n)+p = m+(n+p)$, $(mn)p = m(np)$

Identities

$m+0 = 0+m = m$, $m \cdot 1 = 1 \cdot m = m$

Additive Inverse

For every integer m, there is an integer $-m$ such that $m+(-m) = 0 = (-m)+m$. We call $(-m)$ the additive inverse of m.

Distributive Properties

$m(n+p) = mn+mp$, $(m+n)p = mp+np$

Constructing a Proof for a Conditional Statement

A conditional statement has the form If P then Q and is written $P \to Q$. Usually P is called the hypothesis and Q is called the conclusion. In order to prove that a conditional statement $P \to Q$ is true, we assume that the hypothesis P is true. Then using this assumption along with appropriate definitions, axioms, and theorems, we proceed through a logical sequence of steps to derive the conclusion Q. This method is called a **Direct Proof** and is illustrated in Example 1.

Suppose we explore adding various *odd* integers such as:

$$7 + 9 = 16 \ , \ 5 + 3 = 8 \ , \ 13 + 11 = 24$$

It appears that the sum of any two odd integers is always an even integer. In fact, we would like to prove the following:

" If x and y are odd integers, then $x + y$ is an even integer. "

In order to construct a mathematical proof of this conditional statement, we need precise definitions for the concepts of even and odd integers.

Definition 1 An integer x is an **even integer**, if there exists an integer n such that $x = 2n$. An integer x is an **odd integer**, if there exists an integer m such that $x = 2m + 1$.

Example 1 Prove " if x and y are odd integers, then $x + y$ is an even integer ".

Solution The first step is to clearly identify the hypothesis P and the conclusion Q of the conditional statement that we are trying to prove. In this case, we have

P: x and y are odd integers
Q: $x + y$ is an even integer

We now treat P as what we will assume true and treat Q as what we want to show. Starting with the assumption that P is true, we must make use of it in some constructive way. In this situation, we translate from words into an algebraic form using the definition of "odd integer". Thus, we have the following steps:

Step	Statement	Reason
1	x and y are odd integers	Hypothesis
2	$x = 2n + 1$ and $y = 2m + 1$ for some integers m and n	Definition of odd integer

When x and y are odd integers, we have that $x = 2n + 1$ and $y = 2m + 1$, for some integers m and n . Notice that we do not use $x = 2n + 1$ and $y = 2n + 1$ because that assumes $x = y$. Next, consider the sum $x + y = (2n + 1) + (2m + 1)$. To prove the conclusion, we must show that $x + y$ is even, that is, $x + y = 2q$ for some integer q .
By the associative and commutative properties we obtain

$$x + y = (2n + 1) + (2m + 1) = 2n + 2m + 1 + 1$$

Since $1 + 1 = 2$, this simplifies to

$$x + y = 2n + 2m + 2$$

By the distributive property, this can be written as:

$$x + y = 2n + 2m + 2 = 2(n + m) + 2(1) = 2(n + m + 1)$$

By the closure property, $n + m + 1 = q$ for some integer q and therefore $x + y = 2q$ for some integer q . Consequently, $x + y$ is an even integer. ■

Counterexamples

Sometimes we encounter conditional statements that we want to show are false. This requires that we find a counterexample, that is, we must show that there is a way to make the hypothesis true and the conclusion false. Usually, the hypothesis and conclusion are open sentences, that is, sentences using variables that become true or false statements depending on the values which are substituted for the variables. Thus, to construct a counterexample, we must find values for the variables which make the hypothesis true and the conclusion false.

Example 2 Find a counterexample for the following open sentence:

If x and y are integers and $xy > 0$, then $x > 0$ and $y > 0$.

Solution A counterexample would be given by $x = -3$ and $y = -4$ because then $xy = 12$ and(yet) "$x > 0$ and $y > 0$" is false. ■

Exercises

1. Use the properties of the integers and Definition 1 to prove each of the following:

(a) If x and y are even integers, then $x + y$ is an even integer.

(b) If x is an odd integer and y is even integer, then $x + y$ is an odd integer.

(c) If x is an even integer, then x^2 is an even integer.

(d) If x is an odd integer, then x^2 is an odd integer.

(e) If x and y are even integers, then xy is an even integer.

(f) If x is an odd integer and y is even integer, then xy is an even integer.

(g) If x and y are odd integers, then xy is an odd integer.

2. Find a counterexample for each open sentence:

(a) If $a > b$ and $c > d$, then $ac > bd$.
(b) If x and y are even integers, then $\frac{x}{y}$ is even.

Properties of the Real Numbers

The set of real numbers $\mathbb{R}$ has the arithmetic operations of addition $+$ and multiplication $\cdot$ defined on its elements. We will assume that the following properties are true for addition and multiplication of real numbers a, b, c. These properties are called the Field Axioms.

Closure Properties ✗ Uniqueness?

$a + b$ is a unique real number and $a \cdot b$ (written ab) is a unique real number.

Commutative Properties

$a + b = b + a$, $ab = ba$

Associative Properties

$(a+b)+c = a+(b+c)$, $(ab)c = a(bc)$

Identities

$a+0 = 0+a = a$, $a \cdot 1 = 1 \cdot a = a$

Additive Inverse

For every real number a, there is a real number $-a$ such that $a+(-a) = 0 = (-a)+a$. We call $(-a)$ the additive inverse of a.

Multiplicative Inverse   Does not hold for the set of integers $\mathbb{Z}$

For every real number $a \neq 0$, there is a real number $\frac{1}{a}$ (the multiplicative inverse of a) such that $a(\frac{1}{a}) = 1$. Often, the symbol a^{-1} is used to denote the multiplicative inverse.

Distributive Properties

$a(b+c) = ab+ac$, $(a+b)c = ac+bc$

Observe that as a real number, 2 has a multiplicative inverse $\frac{1}{2}$, since $2(\frac{1}{2}) = 1$. However, $\frac{1}{2}$ is not an integer; thus, the Multiplicative Inverse Property does not hold for the set of integers $\mathbb{Z}$. The set of rational numbers $\mathbb{Q}$ does have the same properties for addition and multiplication as the set of real numbers $\mathbb{R}$. The primary difference between $\mathbb{Q}$ and $\mathbb{R}$ is that $\mathbb{R}$ satisfies the "Completeness Axiom" and contains the irrational numbers. This Axiom will be discussed later. Using only the Field Axioms, many familiar theorems can be proved. Before doing this, we state some basic properties of equality that will be used often.

→ Transitive Property of Equality

For any real numbers a,b,c, if $a = b$ and $b = c$, then $a = c$.

Substitution Property

For any real numbers $a,b,$ if $a = b$, then a may be replaced by b and b may be replaced by a in any mathematical expression without changing the meaning of the expression.

As a consequence of the Substitution Property, we have the following properties.

Addition Property of Equality

For any real numbers a,b,c, if $a = b$, then $a+c = b+c$

Multiplication Property of Equality

For any real numbers a,b,c, if $a = b$, then $ac = bc$.

The Addition Property of Equality means that any real number (or real-valued expression) can be added to both sides of an equation and a true (or equivalent) equation will result.The Multiplication Property of Equality means that any real number (or real-valued expression) can be used to multiply an equation to produce a true (or equivalent) equation.

We now state some theorems, which are used frequently when working with real numbers.

Theorem 1

(i) For any real number a, its additive inverse is unique.
(ii) For any real number a, its multiplicative inverse is unique.

Proof of (i) For any real number a, we know there exists a an additive inverse $-a$ with the property that $a+(-a)=0$. To prove that $-a$ is the unique real number with this property, we must prove that

if b is a real number with $a+b=0$, then $b=-a$.

We present the proof as a sequence of statements that start with the hypothesis that $a+b=0$ and end with the conclusion that $b=-a$. Reasons are provided for getting from one statement to the next.

Statement	Reason
$a+b=0$	assume true
$-a+(a+b)=-a+0$	add $-a$ to each side (to "cancel off" a)
$-a+(a+b)=-a$	property of zero
$(-a+a)+b=-a$	associative property
$0+b=-a$	additive inverse property
$b=-a$	property of zero

The proof of Theorem 1 part (ii) is left as an exercise. ■

Theorem 2

(i) If a is any real number, then $a \cdot 0 = 0 \cdot a = 0$.
(ii) If $ab=0$, then $a=0$ or $b=0$.

Proof of (i) :

Statement	Reason
$a \cdot 0 = a(0+0)$	$0=0+0$ by property of zero
$a \cdot 0 = a \cdot 0 + a \cdot 0$	distributive property
$-(a \cdot 0)+a \cdot 0 = -(a \cdot 0)+a \cdot 0+a \cdot 0$	add $-(a \cdot 0)$ to each side
$0 = 0 + a \cdot 0$	additive inverse property
$0 = a \cdot 0$	property of zero

The proof of (ii) is left as an exercise. ■

Theorem 3 (Properties of Negatives)

(i) $(-1)a=-a$
(ii) $-(-a)=a$
(iii) $(-a)(-b)=ab$
(iv) $(-a)b=a(-b)=-(ab)$

Proof of (i) :

If we can show that $a+(-1)a = 0$, then $(-1)a$ will have the property that the additive inverse $-a$ has. By theorem 1, it will follow that $(-1)a = -a$.

Statement	Reason
$a+(-1)a = (1)\cdot a+(-1)a$	property of 1
$a+(-1)a = (1+-1)a$	distributive property
$a+(-1)a = 0\cdot a$	-1 is the additive inverse of 1
$a+(-1)a = 0$	Theorem 2
$(-1)a = -a$	additive inverse is unique (Theorem 1)

Proof of (iii)

Statement	Reason
$(-a)(-b) = [(-1)a][(-1)b]$	Theorem 3 (i)
$(-a)(-b) = (-1)(-1)ab$	associative and commutative properties
$(-a)(-b) = -(-1)ab$	Theorem 3 (i)
$(-a)(-b) = (1)ab$	$-(-1) = 1$ since $(-1)+1 = 0$
$(-a)(-b) = ab$	property of 1

The proofs of (ii) and (iv) are left as exercises. ■

We can now define subtraction and division as follows.

Definition 2 The subtraction operation $a-b$ is defined by the equation $a-b = a+(-b)$ for all $a,b \in \mathbb{R}$.

Definition 3 For $b \neq 0$, the division of a by b (written $a \div b$) is defined by the equation $a \div b = a\cdot(\frac{1}{b}) = ab^{-1}$. We usually write $a \div b$ as $\frac{a}{b}$.

Theorem 4 (Properties of Division) Assume that $b \neq 0$ and $d \neq 0$.

(i) $\frac{a}{b} = \frac{c}{d}$ if and only if $ad = bc$

(ii) $\frac{a}{b} + \frac{c}{b} = \frac{(a+c)}{b}$

(iii) $(cd)^{-1} = c^{-1}d^{-1}$

(iv) $\frac{a}{b} + \frac{c}{d} = \frac{(ad+bc)}{bd}$

(v) $\frac{ab}{cd} = (\frac{a}{b})(\frac{c}{d})$

Proof of (i) : To prove statement (i) requires proving the following two conditional statements.

(1) If $\frac{a}{b} = \frac{c}{d}$, then $ad = bc$. and (2) If $ad = bc$, then $\frac{a}{b} = \frac{c}{d}$.

The following is a proof for (1):

Statement	Reason
$\frac{a}{b} = \frac{c}{d}$	Assume
$ab^{-1} = cd^{-1}$	definition of division and $\frac{1}{b} = b^{-1}$
$(ab^{-1})b = (cd^{-1})b$	multiply both sides by b
$a(b^{-1}b) = c(d^{-1}b)$	associative property
$a(1) = c(bd^{-1})$	$b^{-1}b = 1$ and commutative
$a = (cb)d^{-1}$	property of 1 and associative
$ad = (cb)d^{-1}d$	multiply both sides by d
$ad = cb(1)$	$dd^{-1} = 1$
$ad = cb$	property of 1
$ad = bc$	commutative

The proof of " If $ad = bc$, then $\frac{a}{b} = \frac{c}{d}$ " can be accomplished by "reversing" the steps in the preceding proof and is left as an exercise.

Proof of (iv) :

The method will be similar to the approach of getting a "common denominator" when adding two fractions. Keep in mind that multiplying a fraction $\frac{a}{b}$ by $\frac{d}{d}$ can be considered as the product $(ab^{-1})(dd^{-1})$.

Statement	Reason
$\frac{a}{b} + \frac{c}{d} = ab^{-1} + cd^{-1}$	definition of division and $\frac{1}{b} = b^{-1}$
$\frac{a}{b} + \frac{c}{d} = ab^{-1}(1) + cd^{-1}(1)$	property of 1
$\frac{a}{b} + \frac{c}{d} = ab^{-1}(dd^{-1}) + cd^{-1}(bb^{-1})$	$dd^{-1} = 1$ and $bb^{-1} = 1$
$\frac{a}{b} + \frac{c}{d} = ad(b^{-1}d^{-1}) + cb(d^{-1}b^{-1})$	commutative and associative
$\frac{a}{b} + \frac{c}{d} = ad(b^{-1}d^{-1}) + bc(b^{-1}d^{-1})$	commutative
$\frac{a}{b} + \frac{c}{d} = ad(bd)^{-1} + cb(bd)^{-1}$	Theorem 4 (iii)
$\frac{a}{b} + \frac{c}{d} = \frac{ad}{bd} + \frac{bc}{bd}$	definition of division and $(bd)^{-1} = \frac{1}{bd}$
$\frac{a}{b} + \frac{c}{d} = \frac{ad+bc}{bd}$	Theorem 4 (ii)

The proofs of (ii), (iii), and (v) are left as exercises. ■

Exercises

1. Prove that if $ab = 1$, then $b = a^{-1}$. (Multiplicative inverse is unique)

2. Prove that if $ab = 0$, then either $a = 0$ or $b = 0$.

3. Prove that $-(-a) = a$.

4. Prove that $(-a)b = a(-b) = -(ab)$.

5. Prove that if $a = -a$, then $a = 0$.

6. Prove that if $ad = bc$, then $\frac{a}{b} = \frac{c}{d}$.

7. Prove that $\frac{a}{b} + \frac{c}{b} = \frac{(a+c)}{b}$.

8. Prove that $(a+b)(c+d) = ac + bc + ad + bd$.

9. Show, by counterexamples, that the operation of subtraction is not commutative.

10. Prove $(-b)^{-1} = -(b^{-1})$ and $\frac{-a}{-b} = \frac{a}{b}$.

11. Prove (a) If $a + b = a + c$, then $b = c$ (b) If $a \neq 0$ and $ab = ac$, then $b = c$.

12. Prove $-(a+b) = -a + -b$.

13. Prove $(cd)^{-1} = c^{-1}d^{-1}$.

14. Prove $\frac{(ab)}{(cd)} = (\frac{a}{c})(\frac{b}{d})$.

15. Show by counterexample that the operation of subtraction is not associative.

16. Prove that $\frac{a}{b} - \frac{c}{b} = \frac{(a-c)}{b}$.

Axioms and Theorems on Inequalities for Real Numbers

We first state some basic axioms and fundamental definitions for the use of $<$ (less than) and $>$ (greater than) operations. Then we use these axioms and definitions to derive many familiar results on inequalities. The set of real numbers consists of the positive real numbers, the negative real numbers, and zero. A real number a is positive means that $0 < a$ and a is negative means that $a < 0$. The following diagram shows the real number line with these features:

_______________________|_______________________

← negative numbers → zero ← positive numbers →

We start with three basic axioms (statements that we assume true without proof) and the definition of " $a > b$ " .

Axiom 1 The positive real numbers are closed under addition $+$ and multiplication $\cdot$.

Axiom 2 If $a > 0$ and $b < 0$, then $ab < 0$ (the product of a positive and negative is a negative).

Definition 4 $a > b$, if there is a positive real number c such that $b + c = a$, that is $a - b > 0$.

Axiom 3 (Trichotomy) For any $a, b \in \mathbb{R}$, exactly one of the following is true:

$$a > b, \quad b > a, \quad \text{or} \quad a = b$$

Theorem 5 (Transitive Property) If $a > b$ and $b > d$, then $a > d$.

Proof:

Statement	Reason
$a > b$ and $b > d$	hypothesis
$b = c_1 + d$ and $a = c_2 + b$ for c_1, c_2 positive real numbers	definition of ">"
$d + (c_1 + c_2) = (d + c_1) + c_2 = b + c_2 = a$	substitution and associative
$c_1 + c_2$ is a positive real number	Axiom 1

Thus, we have shown that $d+$ a positive real number equals a. Consequently, $a > d$. ■

Theorem 6

(a) If $a > 0$, then $a^{-1} > 0$ and $-a < 0$.
(b) If $a < 0$, then $a^{-1} < 0$ and $-a > 0$.

Proof that " if $a > 0$, then $a^{-1} > 0$ "

We will use an indirect proof and show that if $a > 0$, then both the statements $a^{-1} < 0$ and $a^{-1} = 0$ lead to contradictions and therefore can not be true. Then, by the Trichotomy Axiom, $a^{-1} > 0$ will have to be true, since one of $a^{-1} > 0, \; a^{-1} < 0,$ or $a^{-1} = 0$ must be true.
Suppose that $a^{-1} = 0$, then $aa^{-1} = a(0) = 0$; however, $aa^{-1} = 1$ and $1 \neq 0$ so a contradiction results.
Next, suppose that $a^{-1} < 0$, then $a\,a^{-1} < 0$ by Axiom 2, since $a > 0$ and $a^{-1} < 0$. However, $aa^{-1} = 1$ and $1 > 0$ so a contradiction results in this case also.
Thus, $a^{-1} = 0$ and $a^{-1} < 0$ can not be true. Consequently, $a^{-1} > 0$ must be true by Trichotomy.
This completes the proof that " if $a > 0$, then $a^{-1} > 0$."

Proof that " if $a > 0$, then $-a < 0$ "

Again assume that $a > 0$, we will show that $-a < 0$ by a similar indirect proof.
Suppose that $-a > 0$, then $a + (-a) > 0$ by Axiom 1 since both $a > 0$ and $-a > 0$. However, $a + (-a) = 0$ and $0 > 0$ is false so a contradiction results.
Next suppose that $-a = 0$, then $a + (-a) = a + 0 = a > 0$. However, $a + (-a) = 0$ and $0 > 0$ is false so a contradiction results in this case also.
Thus, $-a > 0$ and $-a = 0$ can not be true; hence, $-a < 0$ must be true by Trichotomy.
The proof of (b) is left as an exercise. ■

Corollary 7 If $a < 0$ and $b < 0$, then $(a + b) < 0$.

Proof:

Statement	Reason
$a < 0$ and $b < 0$	assume
$-a > 0$ and $-b > 0$	Theorem 6 (b)
$(-a)+(-b) > 0$	Axiom 1
$-(a+b) > 0$	distributive property: $(-a)+(-b) = -(a+b)$
$-[-(a+b)] < 0$	Theorem 6 (a)
$(a+b) < 0$	$-[-(a+b)] = (a+b)$ by Theorem 3 (ii)

Thus, Corollary 7 follows from Theorem 6. ■

Theorem 8 If $a > b$, then $a+d > b+d$, for any real number d.

Theorem 9 If $a > b$ and $d > 0$, then $ad > bd$ and $\frac{a}{d} > \frac{b}{d}$.

Proof that " if $a > b$ and $d > 0$, then $ad > bd$."

Statement	Reason
$a > b$ and $d > 0$	Assume
$a - b > 0$	definition of "$a > b$"
$(a-b)d > 0$	Axiom 1
$ad - bd > 0$	distributive property
$ad > bd$	definition of "$ad > bd$"

We leave as an exercise the proof that " if $a > b$ and $d > 0$, then $\frac{a}{d} > \frac{b}{d}$ ". ■

Theorem 10 If $a > b$ and $d < 0$, then $ad < bd$ and $\frac{a}{d} < \frac{b}{d}$.
The proofs of Theorems 8 and 10 are left as exercises.

Exercises

1. Prove Theorem 6 part (b).

2. Prove Theorem 8. [Hint: Show $(a+d)-(b+d) > 0$]

3. Prove that If $ad > bd$ and $d > 0$, then $a > b$ [Hint: $d^{-1} > 0$ when $d > 0$]

4. Prove that If $a > b$ and $d > 0$, then $\frac{a}{d} > \frac{b}{d}$. [Hint: $d^{-1} > 0$ when $d > 0$]

5. Prove that If $a > b$ and $d < 0$, then $ad < bd$.

6. Prove that If $a > b$ and $d < 0$, then $\frac{a}{d} < \frac{b}{d}$ [Hint: $d^{-1} < 0$ when $d < 0$]

7. Prove that if $\frac{a}{d} > \frac{b}{d}$ and $d > 0$, then $a > b$.

CHAPTER 2 LOGIC

Introduction

Logic is a comprehensive subject which analyzes statements by using rules for logical operators, devices such as truth tables, and various other techniques. Reasoning patterns are also analyzed.
By a **statement** we mean a declarative sentence or assertion that is either true or false (but not both). For example the sentences:

The natural number 5 is a prime number.
The integer 8 is an even number.

are statements, both of which are true.
Every statement has a truth value, either true (denoted by T) or false (denoted by F). We often use P,Q or R to denote statements, or perhaps $P_1, P_2, \ldots, P_n$ for several statements.
The sentence $P(x) : x^2 = 9$ is an example of an open sentence. $P(x)$ is true, when $x = 3$ and $P(x)$ is false, when $x = 4$. Thus, the truth value of $P(x)$ depends on the value of x. The open sentence $P(a,b,c) : a^2 + b^2 = c^2$ is true, when $a = 4,\ b = 3,\ c = 5$ and false, when $a = 4,\ b = 3,\ c = 10$. In general an **open sentence** is a sentence whose truth value depends on the value(s) of the variable(s) used in the sentence.

Logical Operators and Truth Tables

A **logical operator** (or connective) is a word (or combination of words) that combines one or more statements to make a new "compound" statement. The statements being combined by a logical operator are called "component" statements. For example, the statement

"5 is a prime number and 8 is an even integer"

has the operator "and" joining the component statement "5 is a prime number" with the component statement "8 is an even integer".
The following table gives the names and symbolic forms for some frequently used operators.

Operator	Connective	Sentence Form	Symbolic Form
conjunction	and	P and Q	$P \wedge Q$
disjunction	or	P or Q	$P \vee Q$
negation	not	not P	$\sim P$
conditional	if..then..	if P then Q	$P \rightarrow Q$
biconditional	..if and only if..	P if and only if Q	$P \leftrightarrow Q$

To determine the truth value of a compound statement, we must have a rule for each of the operators "and","or","not","if..then", and " if and only if ". The following definitions describe these rules.

Definition 1 For statements P,Q , the **conjunction** of P with Q (with sentence form "P and Q" and symbolic form $P \wedge Q$) is a statement which is true only in the case that P,Q are both true; otherwise $P \wedge Q$ is false.

Using Definition1, the statement

"5 is a prime number and 8 is an odd number"

must be assigned a truth value of false because the component statement "8 is an odd number" is false. Remember the conjunction $P \wedge Q$ is meant to assert that both P and Q are true.

The possible truth values for a statement can be presented in a table, where each row or line in the table gives a possible combination of truth values for the component statements along with the corresponding value assigned to the statement being analyzed. The next table shows the truth values of $P \wedge Q$ for each of the four possible combinations of truth values of the two component statements P,Q .

P	Q	$P \wedge Q$
T	T	T
T	F	F
F	T	F
F	F	F

Table 1

Definition 2 For statements P,Q , the **disjunction** of P with Q (with sentence form "P or Q" and symbolic form $P \vee Q$) is a statement which is false only in the case that P,Q are both false; otherwise $P \vee Q$ is true.

According to Definition 2, the statement

"4 is an odd number or 5 is a prime number"

is a true statement because one of its components; namely, "5 is a prime number" is true. Intuitively, the disjunction $P \vee Q$ is the assertion that one or the other of P,Q is true. In mathematics, "or" is used in the inclusive sense, that is, it includes the situation where both components are true.

The following truth table shows the truth values of $P \vee Q$ for each of the four possible combinations of truth values of the two component statements P,Q .

P	Q	$P \vee Q$
T	T	T
T	F	T
F	T	T
F	F	F

Table 2

Definition 3 For a statement P , the **negation** of P (with sentence form "it is false that P" and symbolic form $\sim P$) is the statement which is true when P is false and false when P is true.

According to Definition 3, the statement

"It is false that 4 is an even number"

is a false statement because it is the negation of the statement "4 is an even number" , which is a true statement. Keep in mind that the negation $\sim P$ is stating the opposite of P .

The following truth table shows the truth values of $\sim P$ for each of the possible values of P:

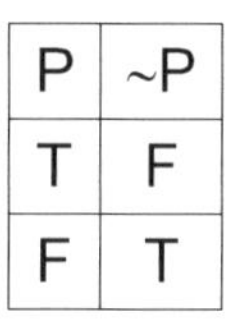

P	~P
T	F
F	T

Table 3

Conditional statements are very important in mathematics because almost all mathematical theorems can be posed in the form of a conditional statement. Most theorems have the form

If hypothesis, then conclusion.

The following is an illustrative example:

If two lines in the plane are perpendicular to the same line, then they are parallel lines.

Definition 4 For statements P,Q , the **conditional** (with sentence form "if P then Q" and symbolic form $P \rightarrow Q$) is a statement which is false only in the case that P is true and Q is false; otherwise $P \rightarrow Q$ is true. P is called the hypothesis of the conditional $P \rightarrow Q$ and Q is called the conclusion.

According to Definition 4, the statement

If x is an integer with $x < 3$, then $x^2 < 9$.

is false when $x = -4$ and true when $x = 2$.

Notice that in this example, we are technically forming the conditional using two open sentences (not statements), However, when values are substituted for the variables, we have statements and the rule for the conditional can be applied.

The following table shows the truth values of $P \rightarrow Q$ for each of the four possible combinations of truth values of the two component statements P,Q .

P	Q	P → Q
T	T	T
T	F	F
F	T	T
F	F	T

Table 4

In the English language, there are several ways to express the conditional statement $P \rightarrow Q$ other than the form " If P then Q ".This can be illustrated as follows:

Form	Example
If P then Q	If x is even, then x is a multiple of 2 .
P implies Q	x is even implies x is a multiple of 2 .
P only if Q	x is even only if x is a multiple of 2 .
Q is a necessary condition for P	x is a multiple of 2 is necessary for x to be even.
P is a sufficient condition for Q	x is even is sufficient for x to be a multiple of 2 .

Truth Table Methods and Equivalent Statements

Suppose we are interested in determining whether the negation of the conjunction $P \wedge Q$ (written $\sim(P \wedge Q)$) is given by $\sim P \wedge \sim Q$ or given by $\sim P \vee \sim Q$. If we need to find the negation of:

" x is odd and x is prime. "

Then, the choice $\sim P \wedge \sim Q$ gives " x is not odd and x is not prime ".

While the choice $\sim P \vee \sim Q$ gives " x is not odd or x is not prime ". We need to determine which is the correct choice.

We can solve this problem by comparing truth tables for the statements $\sim(P \wedge Q)$, $\sim P \wedge \sim Q$, and $\sim P \vee \sim Q$.

The component method is used to construct the table for $\sim(P \wedge Q)$, as follows:

P	Q	$P \wedge Q$	$\sim(P \wedge Q)$
T	T	T	F
T	F	F	T
F	T	F	T
F	F	F	T

Table 5

The values in the column for $P \wedge Q$ are created by applying the rule for the "and" operator to the values of P,Q in each row. Then, the values in the column for $\sim(P \wedge Q)$ result from applying the "negation" rule to the values of $P \wedge Q$ in each row.

The "Step Method" is used for $\sim P \wedge \sim Q$ as follows:

	P	Q	$\sim P$	$\wedge$	$\sim Q$
	T	T	F		F
	T	F	F		T
	F	T	T		F
	F	F	T		T
Step No.	1	1	2		3

Table 6-a

In Step 1, we assign all the possible values to the components P,Q . Then in Step 2, we assign values to ~P using the negation rule applied to the values of P in each row, and in Step 3, we assign values to ~Q by applying the negation rule to the values of Q in each row. Table 6-a shows the resulting table.
Finally, we complete Table 6-a to obtain Table 6-b by Step 4, which assigns values to $\sim P \wedge \sim Q$ using the rule for the operator "and" applied to the values of ~P and ~Q in each row of Table 6-a.

	P	Q	~P	∧	~Q
	T	T	F	F	F
	T	F	F	F	T
	F	T	T	F	F
	F	F	T	T	T
Step No.	1	1	2	4	3

Table 6-b

Observe that the final set of values for $\sim P \wedge \sim Q$ (determined by Step 4) is *not* the same pattern of truth values for the four rows shown in Table 5 for the statement $\sim(P \wedge Q)$.
In a similar fashion, we use the Step Method to create a Table for $\sim P \vee \sim Q$ as follows:

	P	Q	~P	∨	~Q
	T	T	F	F	F
	T	F	F	T	T
	F	T	T	T	F
	F	F	T	T	T
Step No.	1	1	2	4	3

Table 7

For Step 4 in Table 7, we assign values to $\sim P \vee \sim Q$ by using the rule for the "or" operator applied to the values of ~P and to the values of ~Q in each row of the table.
Observe that the final set of values for $\sim P \vee \sim Q$ (determined by Step 4) is exactly the same pattern of truth values for the four rows shown in Table 5 for the statement $\sim(P \wedge Q)$.
We say that $\sim(P \wedge Q)$ is logically equivalent to $\sim P \vee \sim Q$ and write $\sim(P \wedge Q) \approxeq \sim P \vee \sim Q$. We formalize this concept by the following definition

Definition 5 Two statements are **logically equivalent** provided that they both have exactly the same pattern of truth values assigned for all possible combinations of truth values of their component statements. The symbol $\approxeq$ is used to denote logical equivalence.

We have verified the first of DeMorgan's Laws, which can be stated as follows:

Theorem 1 (DeMorgan's Laws)
(1) $\sim(P \wedge Q) \approxeq \sim P \vee \sim Q$
(2) $\sim(P \vee Q) \approxeq \sim P \wedge \sim Q$

The verification of Law (2) is left as an exercise. ■

We can now use DeMorgan's First Law to determine that the negation of

" x is odd and x is prime "

is given by the statement:

" x is not odd or x is not prime "

Example 1 Verify that $\sim (\sim P) \approxeq P$.

Solution The following table shows that $\sim (\sim P)$ has the same truth values as P in all cases.

P	~P	~ (~P)
T	F	T
F	T	F

Table 8

Variations of the Conditional

Given any conditional statement of the form " If hypothesis, then conclusion " , mathematicians are naturally interested in determining the truth value of its converse, which has the form " If conclusion, then hypothesis " .

Definition 6 For statements P,Q , the **converse** of $P \rightarrow Q$ is the conditional $Q \rightarrow P$.

Consider the statement

"If a four-sided figure is a square, then it is a rectangle."

This statement is a basic fact of plane geometry. Its converse is the following statement.

"If a four-sided figure is a rectangle, then it is a square."

The converse is false because not all rectangles have four equal sides and hence not all rectangles are squares. This example illustrates the fact that the converse and the original conditional may have opposite truth values.

Consider the statement:

"If n^2 is odd, then n is odd."

This statement is difficult to prove directly. Instead, we can formulate its contrapositive (using the next definition) and obtain:

"If n is not odd, then n^2 is not odd."

Using the fact that an integer is either even or odd, this can be written as:

"If n is even, then n^2 is even."

This statement is called the contrapositive of the original statement and is easy to prove. Moreover, its proof in effect proves the original statement because by Theorem 2 (which follows), a statement and its contrapositive are equivalent.

Definition 7 For statements P,Q , the **contrapositive** of the conditional $P \to Q$ is the conditional $\sim Q \to \sim P$.

Theorem 2 The contrapositive $\sim Q \to \sim P$ is logically equivalent to the conditional $P \to Q$.

Proof Construct a truth table for $\sim Q \to \sim P$ as follows:

	P	Q	~Q	→	~P
	T	T	F	T	F
	T	F	T	F	F
	F	T	F	T	T
	F	F	T	T	T
Step No.	1	1	2	4	3

Table 9

In Table 9, Step 4 assigns values to $\sim Q \to \sim P$ by using the rule for " → " applied to the values of ~Q and ~P in each row. Table 9 shows that the pattern of assigning truth values to $\sim Q \to \sim P$ is the same as the pattern of assigning values to $P \to Q$ for all possible combinations of truth values for P,Q as shown in Table 4. Thus, it follows that $\sim Q \to \sim P$ is logically equivalent to $P \to Q$. ■

The significance of Theorem 2 is that we can establish whether $P \to Q$ is true (or false) by establishing that $\sim Q \to \sim P$ is true (or false). In other words, since a conditional statement and its contrapositive are equivalent, then we can work with either statement to determine its truth value.

Theorem 3 The conditional statement $P \to Q$ is logically equivalent to $\sim P \vee Q$.

The proof of Theorem 3 is left as an exercise.

One of the primary goals of logic is to formulate the negation of any given statement even though it may be a compound statement. For example, DeMorgan's Laws give methods for negating conjunctions and disjunctions. The next theorem shows how to negate a conditional statement.

Theorem 4 The **negation of the conditional statement** $P \to Q$ is logically equivalent to the statement $P \wedge \sim Q$, that is, $\sim (P \to Q) \cong P \wedge \sim Q$.

Proof Theorem 4 can be proved using truth tables (see Exercise 9). However, the following method uses what is called "Propositional Logic":

By Theorem 3, $P \to Q \approxeq \sim P \vee Q$; thus,

$\sim(P \to Q) \approxeq \sim(\sim P \vee Q)$

Then by DeMorgan's Laws and the result that $\sim(\sim P) \approxeq P$, we have

$$\sim(\sim P \vee Q) \approxeq \sim(\sim P) \wedge \sim Q \approxeq P \wedge \sim Q \quad \blacksquare$$

Theorem 4 shows that the negation of a conditional statement is not another conditional statement, but instead a conjunction. Moreover, when we want a counterexample for the statement $P \to Q$, we must find an example that makes P true and Q false. Therefore, we must make P true and $\sim Q$ true, that is, we must make the statement $P \wedge \sim Q$ true. Thus, when we find a counterexample to the conditional $P \to Q$, then we have an example that makes its negation $P \wedge \sim Q$ true.

Example 2 Find the negation of "if x is even, then x is not prime" and a value for x that makes the negation true.

Solution: This has the form $P \to Q$, where P is "x is even" and Q is "x is not prime" .

By Theorem 4, the negation has the form $P \wedge \sim Q$, which states

"x is even and it is false that x is not prime"

This can be simplified to

"x is even and x is prime"

This statement is true when $x = 2$. Thus, 2 gives a counterexample to the original statement. ■

Sometimes when we are attempting to prove a theorem, we may not be able to develop a proof for the original statement of the theorem. However, in some cases, we may be able to prove an equivalent statement. Since the two statements are equivalent, the proof of one is in effect a proof of the other. This is illustrated in the following example.

Example 3 Formulate a statement equivalent to "If xy is even, then x is even or y is even" .

Solution We use the fact that the contrapositive is equivalent. Since the given statement has the form $P \to (Q \text{ or } R)$, the contrapositive has the form $\sim(Q \text{ or } R) \to \sim P$. By DeMorgan's Laws, this is equivalent to $(\sim Q \text{ and } \sim R) \to \sim P$.

In words, we obtain

"If x is not even and y is not even, then xy is not even."

This can be simplified to:

"If x is odd and y is odd, then xy is odd."

Recall that this statement has been proven previously. ■

Definition 8 For statements P,Q , the **biconditional** (with sentence form "P if and only if Q" and symbolic form $P \leftrightarrow Q$) is the conjunction " $P \to Q$ and $Q \to P$ " . In other words, we have the equivalence $P \leftrightarrow Q \approxeq (P \to Q) \wedge (Q \to P)$.

Example 4 Use the component method to construct a truth table for $(P \to Q) \wedge (Q \to P)$:

Solution Using the rules for "$\to$" and "$\wedge$" , we can create the following table:

P	Q	$P \to Q$	$Q \to P$	$(P \to Q) \wedge (Q \to P)$
T	T	T	T	T
T	F	F	T	F
F	T	T	F	F
F	F	T	T	T

Table 10

Here, we use the abstract rule that (left) → (right) is false only when (left) is true and (right) is false. ■

Example 4 shows that the biconditional is true only in the cases where both P,Q are true or both P,Q are false.

In mathematics, we often must prove a biconditional statement $P \leftrightarrow Q$. This requires that we prove the conditional statement $P \to Q$ and also prove its converse $Q \to P$. For example, to prove the statement:

" n^2 is even if and only if n is even "

requires proving the two statements:

(1) If n^2 is even, then n is even. and (2) If n is even, then n^2 is even.

The following chart states some of the most frequently used logical equivalences. We have already verified many of these equivalences.

Summary of Useful Equivalences

DeMorgan's Laws

$$\sim(P \wedge Q) \approxeq \sim P \vee \sim Q$$
$$\sim(P \vee Q) \approxeq \sim P \wedge \sim Q$$

Conditional Statements

$$P \to Q \approxeq \sim Q \to \sim P \quad \text{(contrapositive)}$$
$$P \to Q \approxeq \sim P \vee Q$$
$$\sim(P \to Q) \approxeq P \wedge \sim Q \quad \text{(negation)}$$

Biconditional Statement

$$P \leftrightarrow Q \approxeq (P \to Q) \wedge (Q \to P) .$$

Double Negation

$$\sim(\sim P) \approxeq P$$

Distributive Laws

$$P \vee (Q \wedge R) \approxeq (P \vee Q) \wedge (P \vee R)$$
$$P \wedge (Q \vee R) \approxeq (P \wedge Q) \vee (P \wedge R)$$

Conditionals

$$P \to (Q \vee R) \approxeq (P \wedge \sim Q) \to R$$
$$(P \vee Q) \to R \approxeq (P \to R) \vee (Q \to R)$$
$$P \to (Q \wedge R) \approxeq (P \to Q) \wedge (P \to R)$$
$$P \to (Q \to R) \approxeq (P \wedge Q) \to R$$

Example 5 Use Truth Tables to show that $P \to (Q \vee R) \approxeq (P \wedge \sim Q) \to R$.

Solution For three statements, P,Q,R , there are eight possible combinations of truth values that can be assigned. In the following table, Step 1 assigns these values. To determine the corresponding values of $P \to (Q \vee R)$, we find the values of P in Step 2 and use Step 3 to find the values of $(Q \vee R)$. Then, in Step 4, the rule for the implies "$\to$" is applied row by row to the values of P and $(Q \vee R)$. The resulting values are displayed in the column above the "4" in the table. The table is then extended to show values for $(P \wedge \sim Q) \to R$. This requires 5 steps labeled Step No. $2'$, $3'$, $4'$, $5'$, and $6'$.
We must use the rule for "$\to$" abstractly and realize that (left) $\to$ (right) is false only when (left) is true and (right) is false.

	P	Q	R	P	$\to$	$(Q \vee R)$	$(P$	$\wedge$	$\sim Q)$	$\to$	R
	T	T	T	T	T	T	T	F	F	T	T
	T	T	F	T	T	T	T	F	F	T	F
	T	F	T	T	T	T	T	T	T	T	T
	T	F	F	T	F	F	T	T	T	F	F
	F	T	T	F	T	T	F	F	F	T	T
	F	T	F	F	T	T	F	F	F	T	F
	F	F	T	F	T	T	F	F	T	T	T
	F	F	F	F	T	F	F	F	T	T	F
Step No.	1	1	1	2	4	3	$2'$	$4'$	$3'$	$6'$	$5'$

Table 11

The pattern in column 4 gives the truth values for $P \to (Q \vee R)$, while the pattern in column $6'$ gives the truth values for $(P \wedge \sim Q) \to R$. Since these patterns are the same for all possible combinations of truth values for P,Q,R , we have verified that $P \to (Q \vee R) \approxeq (P \wedge \sim Q) \to R$. ■

The proofs of those equivalences listed in the Summary of Useful Equivalences that were not previously verified are left as exercises.

Exercises

1. Assume that each component has the obvious truth value (for example, "14 is a prime number" is false). Decide the truth value (T or F) of each of the given statements.

(a) 16 is an even number and 9 is a prime number

(b) If $x = 2$, then $x^2 = 4$.

(c) It is false that $2 + 2 = 5$.

(d) 15 is a prime number or 15 is an odd number

2. Construct a truth table for each of the following statements:

(a) $P \vee \sim Q$ (b) $(P \vee Q) \wedge \sim Q$ (c) $\sim(P \vee Q) \wedge \sim Q$

3. Construct an eight-line table (see Table 11) for each of the following:

(a) $(P \wedge Q) \rightarrow R$ (b) $(P \vee Q) \rightarrow \sim R$

4. Use truth tables to determine if the statements in each pair are logically equivalent.

(a) $P \rightarrow Q$ and $\sim(P \wedge \sim Q)$ (b) $P \vee Q$ and $\sim P \rightarrow Q$

5. Use DeMorgan's Laws to write the negation of each of the following:

(a) x is even and x is prime

(b) $x > 2$ or $x < -5$

(c) $x \neq 4$ or $x^2 < 5$

6. Given " If x and y are even, then xy is even ". Write statements for each:

(a) converse (b) contrapositive (c) negation

7. Use truth tables to verify Theorem 1 part (2).

8. Use truth tables to verify Theorem 3.

9. Use truth tables to verify that $\sim (P \rightarrow Q) \approxeq P \wedge \sim Q$.

10. Use an eight-line truth table to verify that $P \vee (Q \wedge R) \approxeq (P \vee Q) \wedge (P \vee R)$.

11. Let P represent " x is even " and Q represent " x^2 is even " . Express the conditional statement $P \rightarrow Q$ as an English sentence using:

(a) the " if..then.." form of the conditional

(b) the word " implies "

(c) the " only if " form of the conditional

(d) the phrase " is necessary for "

(e) the phrase " is sufficient for "

12. Find the negation of the conditional " if x is a multiple of π , then $\cos(x) = 1$ ". Then find a value of x which makes the negation true. Is this a counterexample to the given conditional?

13. For $a \in \mathbb{R}$, consider the statement " if $a > 0$, then $a^{-1} > 0$ ".
(a) write the contrapositive (b) write the negation

14. Use an eight-line truth table to verify that P → (Q ∧ R) ≊ (P → Q) ∧ (P → R) .

15. The conditional ~P → ~Q is called the **inverse** of P → Q . Show that ~P → ~Q is not logically equivalent to P → Q .

16. Consider the open sentences

$$\text{P}(x)\text{: } x = -3 \text{ and Q}(x)\text{: } x^2 = 9$$

State each of the following in words:
(a) P → Q (b) Q → P (c) P ↔ Q

17. Given that (Q ∨ R) → ~P is false and Q is false, find the truth values of R and P .

18. If W,X,Y,Z are four distinct statements, how many possible combinations of truth values can be assigned to W,X,Y,Z ?

19. The statement [(P → Q) ∧ (Q → R)] → (P → R) is called the Law of Hypothetical Syllogism. Set up an eight-line truth table to show that this statement is true for each line of its truth table.

20. Use truth tables to verify that P → (Q → R) ≊ (P ∧ Q) → R .

CHAPTER 3 PROOFS ABOUT INTEGERS

Multiples and Divisors

We begin with two basic definitions.

Definition 1 For $x,y \in \mathbb{Z}$, with $x \neq 0$, x **is a divisor of** y (written $x \mid y$) provided that there is some integer n such that $xn = y$. Zero is not a divisor of any integer.

If x is not a divisor of y, then we write $x \nmid y$. Observe that $3 \mid 12$ since $3 \cdot 4 = 12$; however, $5 \nmid 12$ because $5n \neq 12$ for all $n \in \mathbb{Z}$.

Definition 2 For any $n \in \mathbb{Z}$, the set of all multiples of n is

$$\{n \cdot t \mid t \in \mathbb{Z}\}$$

The set of multiples of n is usually written $n\mathbb{Z}$ and each member of this set is called a multiple of n. For example,

$$3\mathbb{Z} = \{\ldots -9, -6, -3, 0, 3, 6, 9, \ldots\}$$

Note that the statements "x is a divisor of y" and "y is a multiple of x" are synonymous. Also, it is true that 1 is a divisor of every integer y, since $1 \cdot y = y$ is true for any $y \in \mathbb{Z}$.

Definition 3 For $p \in \mathbb{N}$, with $p > 1$, p **is prime** means the only natural numbers which are divisors of p are 1 and p itself. A natural number that is not prime is a **composite number**.

The first several primes are $2, 3, 5, 7, 11, 13, 17,$ and 19.

Consider the set $3\mathbb{Z} = \{\ldots -9, -6, -3, 0, 3, 6, 9, \ldots\}$, it appears that if $x,y \in 3\mathbb{Z}$, then $(x+y) \in 3\mathbb{Z}$ so that $3\mathbb{Z}$ is closed under addition. This can be verified as follows: Suppose that $x,y \in 3\mathbb{Z}$, then $x = 3n$ and $y = 3m$ for some $m,n \in \mathbb{Z}$ Thus

$$x + y = 3n + 3m = 3(n+m) = 3q$$

Here, $q = n + m$ is an integer, since $\mathbb{Z}$ is closed under addition and thus $x+y$ is a multiple of 3. We will be able to prove that for any integer n, the set $n\mathbb{Z}$ is closed under addition. The next theorem provides the basis for proving this.

Theorem 1 If $a \mid b$ and $a \mid c$, then $a \mid (b+c)$ and $a \mid (b-c)$.

Proof: Assume $a \mid b$ and $a \mid c$, then there are integers m,n such that $am = b$ and $an = c$. Therefore,

$$am + an = b + c$$

By the distributive property, $am + an = a(m+n)$ and since $\mathbb{Z}$ is closed under addition, there is an integer q, with $q = m + n$. Thus,

$$b + c = am + an = a(m+n) = aq$$

Consequently, $a \mid (b+c)$.

The proof that " If $a \mid b$ and $a \mid c$, then $a \mid (b-c)$." is left as an exercise. ■

Corollary 2 For any integer n, the set $n\mathbb{Z}$ is closed under addition and subtraction.

The next theorem has several useful consequences.

Theorem 3 If $a \mid b$ and $b \mid c$, then $a \mid c$.

Proof: To prove that $a \mid c$, we must find an integer q such that $aq = c$. Because $a \mid b$ and $b \mid c$, we know there are integers m, n such that $am = b$ and $bn = c$. By substitution, we obtain $(am)n = c$. Thus, $a(mn) = c$ and we can use $q = mn$, which is an integer, since $\mathbb{Z}$ is closed for multiplication. This shows that $aq = c$ and hence $a \mid c$. ■

Corollary 4 Any multiple of an even number is an even number.

Corollary 5 If $a \mid b$ and c is a multiple of b, then $a \mid c$.

The proofs of these corollaries are left as exercises.

Definition 4 For $x, y \in \mathbb{Z}$, t is a **common divisor** of x and y means $t \mid x$ and $t \mid y$.

Note that 12 is a common divisor of 72 and 144. Also, 3 is a common divisor of 9 and 15.

Theorem 6 If $a, x \in \mathbb{N}$ and $a \mid x$, then $a \leq x$.

Proof: Because $a, x \in \mathbb{N}$, we have that $1 \leq a$ and $1 \leq x$. Moreover, since $a \mid x$, there is an integer n such that $an = x$. We claim that $n > 0$ because if $n \leq 0$, then multiplying both sides of the inequality $n \leq 0$ by $a > 0$ gives $an \leq a(0)$, that is, $an \leq 0$. However, this makes $x \leq 0$ since $x = an$ and thereby contradicts that $1 \leq x$.Therefore, n is a positive integer and consequently $1 \leq n$. Thus, multiplying both sides of the inequality $1 \leq n$ by $a > 0$, we obtain that $a = a(1) \leq an = x$, that is, $a \leq x$. ■

Theorem 6 proves that a divisor of a positive integer can not be larger than that integer itself.Thus, if at least one of two integers is not zero, then a common divisor can not be larger than the absolute value of the nonzero integer. Thus, any pair of integers, which are not both zero, can only have a finite number of common divisors. Thus, there must be a largest common divisor.

Definition 5 For $x, y \in \mathbb{Z}$, t is the **greatest common divisor** provided that
(i) t is a common divisor of x and y; and
(ii) every common divisor is less than or equal to t.

The greatest common divisor of x and y is written GCD(x,y).
For example, GCD$(18,27) = 9$ and GCD$(8,15) = 1$.
Note that if $x = 0$ and $y = 0$, then there is no greatest common divisor because every nonzero integer is a common divisor of zero.

A general problem is the following:

Given $a, b, c \in \mathbb{Z}$, when does $ax + by = c$ have integral solutions x and y?

The next theorem gives a criteria that specifies when $ax + by = c$ can not have integral solutions.

Theorem 7 For $a, b, c \in \mathbb{Z}$ and $d \in \mathbb{Z} - \{0\}$, if $d \mid a$ and $d \mid b$ and $d \nmid c$, then the equation $ax + by = c$ has no integral solution for x and y.

Proof: Suppose that $d \mid a$ and $d \mid b$. Then, for any integers x and y, Corollary 5 shows that $d \mid (ax)$ and $d \mid (by)$. Consequently, $d \mid (ax + by)$ from Theorem 1. Therefore, if there are integers x and y such that $ax + by = c$, then it follows that $d \mid c$, since $d \mid (ax + by)$. However, this would contradict the assumption that $d \nmid c$. Thus, $ax + by = c$ can not have integral solutions for x and y. ■

Theorem 7 shows, for example, that the equation $3x + 15y = 7$ has no integral solutions since $3 \mid 3$ and $3 \mid 15$ and $3 \nmid 7$.

For many pairs of integers x and y, the only positive common divisor is 1 and in this case GCD$(x, y) = 1$.

Definition 6 For $x, y \in \mathbb{Z}$, x and y are **relatively prime** provided that GCD$(x, y) = 1$.

For example, GCD$(8, 15) = 1$ means that 8 and 15 are relatively prime even though both 8 and 15 are not prime.

Congruence Modulo n

We now define when two integers are congruent modulo n for some natural number n.

Definition 7 Let $n \in \mathbb{N}$ and $a, b \in \mathbb{Z}$, then a **is congruent to** b **modulo** n provided that $n \mid (a - b)$. A standard notation for "a is congruent to b modulo n" is $a \approxeq b(\text{mod}\, n)$.

Notice that $a \approxeq b(\text{mod}\, n)$ means there exists $k \in \mathbb{Z}$ such that $nk = (a - b)$ or $a = b + nk$.

Example 1 Let $n = 3$. Find the set $\{x \mid x \approxeq 1(\text{mod}\, 3)\}$.

Solution: It can be verified that $4 \approxeq 1(\text{mod}\, 3)$, $7 \approxeq 1(\text{mod}\, 3)$, $10 \approxeq 1(\text{mod}\, 3)$, and so forth. In fact, the set

$$A = \{\ldots -5, -2, 1, 4, 7, 10, \ldots\} = \{x \mid x \approxeq 1(\text{mod}\, 3)\}$$

We create A by starting with the number 1 and adding all the possible multiples of 3 to it, because $x \approxeq 1(\text{mod}\, 3)$ if and only if $x = 1 + 3k$ for $k \in \mathbb{Z}$. ■

We can also show that

$$B = \{x \mid x \approxeq 2(\text{mod}\, 3)\} = \{\ldots -4, -1, 2, 5, 8, \ldots\}$$

Similarly, the set

$$C = \{x \approxeq 0(\bmod 3)\} = \{\ldots -6, -3, 0, 3, 6, \ldots\}$$

Now observe that $A \cup B \cup C = \mathbb{Z}$ and $A \cap B = \varnothing$, $A \cap C = \varnothing$, and $B \cap C = \varnothing$. We say that the sets A, B, C form a partition of $\mathbb{Z}$. We will generalize these ideas when we study equivalence relations.

Theorem 8 (Properties of Congruence Modulo n)

Let $n \in \mathbb{N}$ and $a, b, c \in \mathbb{Z}$, then

(1) For any integer a, $a \approxeq a(\bmod n)$;

(2) If $a \approxeq b(\bmod n)$, then $b \approxeq a(\bmod n)$;

(3) If $a \approxeq b(\bmod n)$ and $b \approxeq c(\bmod n)$, then $a \approxeq c(\bmod n)$.

Proof of (2): Suppose that $a \approxeq b(\bmod n)$ then $(a-b) = nk$ for some $k \in \mathbb{Z}$. Now

$$(b-a) = -(a-b) = -(nk) = n(-k)$$

Because $-k \in \mathbb{Z}$, whenever $k \in \mathbb{Z}$, this equation shows that $n(-k) = (b-a)$ and hence $n \mid (b-a)$. Consequently by the definition of " congruence modulo n ", we have shown that $b \approxeq a(\bmod n)$, when $a \approxeq b(\bmod n)$.

Proof of (3): Suppose that $a \approxeq b(\bmod n)$ and $b \approxeq c(\bmod n)$. Then, there exist integers k and q such that $a-b = nk$ and $b-c = nq$. By adding these two equations, we obtain

$$(a-b)+(b-c) = nk+nq$$

Simplifying this equation gives

$$(a-c) = nk+nq = n(k+q)$$

By the closure property of $\mathbb{Z}$ under addition, $(k+q) \in \mathbb{Z}$. Thus, $n \mid (a-c)$ and hence $a \approxeq c(\bmod n)$. The proof of (1) is left as an exercise. ■

Property (1) is called the reflexive property, property (2) is called the symmetric property, and property (3) is called the transitive property of congruence modulo n.

Theorem 9 Let $n \in \mathbb{N}$ and $a, b, c, d \in \mathbb{Z}$. If $a \approxeq b(\bmod n)$ and $c \approxeq d(\bmod n)$, then

(1) $(a+c) \approxeq (b+d)(\bmod n)$

(2) $ac \approxeq bd(\bmod n)$

(3) For $m \in \mathbb{N}$, $a^m \approxeq b^m(\bmod n)$

The proof of (1) is left as an exercise.

Proof of (2): Suppose that $a \approxeq b(\bmod n)$ and $c \approxeq d(\bmod n)$, then for some $k, q \in \mathbb{Z}$ we have $a = b+nk$ and $c = d+nq$. Thus,

$$ac = (b+nk)(d+nq) = bd+bnq+dnk+n^2kq = bd+n(bq+dk+nkq)$$

By subtracting bd from both sides of the above equation, we obtain

$$ac-bd = n(bq+dk+nkq)$$

Since $bq+dk+nkq$ is an integer, this proves that $n \mid (ac-bd)$ and hence $ac \approxeq bd(\bmod n)$.

Proof of (3): From part (2), since $a \approx b(\bmod n)$ and $a \approx b(\bmod n)$, we obtain

$$a(a) \approx b(b)(\bmod n) \text{, that is, } a^2 \approx b^2(\bmod n)$$

Continuing with $a \approx b(\bmod n)$ and $a^2 \approx b^2(\bmod n)$, we obtain using (2)

$$a(a^2) \approx b(b^2)(\bmod n) \text{, that is } a^3 \approx b^3(\bmod n) .$$

Similarly, $a^4 \approx b^4(\bmod n)$.

However, to construct a formal proof that $a^m \approx b^m(\bmod n)$ for any natural number m, requires the Principle of Mathematical Induction. This will be studied later. ■

Theorem 9 provides tools to explore certain properties of natural numbers. For example, since $3^4 = 81$, we can conclude that $3^4 - 1 = 80$ and hence that $3^4 \approx 1(\bmod 10)$. From Theorem 9, then $3^4 3^4 \approx 1(1)(\bmod 10)$. Thus, $3^8 \approx 1(\bmod 10)$. Consequently, $3^8 - 1$ is a multiple of 10. This tells us that the last digit in the decimal representation of 3^8 is a 1.

The following theorem can also be proved.

Theorem 10 Let $n \in \mathbb{N}$ and $a, b \in \mathbb{Z}$. If $a \approx b(\bmod n)$, then $ka \approx kb(\bmod n)$ for any $k \in \mathbb{Z}$.

Proof: Suppose that $a \approx b(\bmod n)$, then $a - b = nq$ for some $q \in \mathbb{Z}$. Therefore,

$$k(a-b) = k(nq) \text{, that is, } ka - kb = n(kq) .$$

Since $kq \in \mathbb{Z}$ by closure for multiplication, $n \mid (ka - kb)$ and hence $ka \approx kb(\bmod n)$. ■

Further Results for Even and Odd Integers

We now analyze the question of whether there exists an integer that is both even and odd.

Theorem 11 If $u \in \mathbb{Z}$, then $2u \neq 1$.

Proof: We consider the three possible cases:

$$u < 0, \; u = 0, \text{ or } u > 0$$

Case 1 $u < 0$

Since $2 > 0$, we have from inequality properties that $2u < 2 \cdot 0 = 0$. However, $1 > 0$, thus $2u \neq 1$.

Case 2 $u = 0$

If $u = 0$, then $2u = 2 \cdot 0 = 0$. Since $1 \neq 0$, we have that $2u \neq 1$ in this case also.

Case 3 $u > 0$

Here $u > 0$ and $u \in \mathbb{Z}$ mean that $u \geq 1$. Therefore, multiplying both sides of the inequality $u \geq 1$ by 2 and using the fact that $2 > 0$, we obtain that $2u \geq 2(1) = 2$, that is, $2u \geq 2$. Since $2 > 1$, we have $2u \neq 1$ in this case also.

Thus, for every possible value of $u \in \mathbb{Z}$, we have that $2u \neq 1$.

Theorem 12 For $x \in \mathbb{Z}$, if x is odd, then x is not even.

Proof: Suppose that x is odd and x is also even. Then there are integers $m, n \in \mathbb{Z}$ such that $x = 2m + 1$ and $x = 2n$. Therefore, $2m + 1 = 2n$. This equation can be written as

$$1 = 2n - 2m = 2(n - m)$$

Letting $u = (n - m)$, we have that $u \in \mathbb{Z}$, since $\mathbb{Z}$ is closed for addition and $(n - m) = n + (-m)$. Thus, we have found a $u \in \mathbb{Z}$ such that $2u = 1$, that is , we have a counterexample to theorem 11. However, Theorem 11 is true and can not have any counterexamples. Consequently, we can not have an integer x that is both even and odd. ■

Theorem 12 does not prove that an integer is either even or odd because it is still possible that an integer could be neither even nor odd. The formal proof that this is impossible is based on the Division Algorithm, which will be discussed later.

From now on, we will assume that every integer must be either even or odd and can not be both even and odd. This concept will enable us to prove results by the method of contrapositives or the contradiction method. This is illustrated by the next example.

Example 2 Prove that if n^2 is even, then n is even.

Solution A direct proof would assume that n^2 is even and derive that n is even. However, knowing that $n^2 = 2q$ for some $q \in \mathbb{Z}$ does not lead to $n = 2p$ for some $p \in \mathbb{Z}$. From our study of logic, we know that a conditional statement $P \to Q$ and its contrapositive $\sim Q \to \sim P$ are equivalent. Thus, we could prove the contrapositive of " if n^2 is even, then n is even " . Its contrapositive states " if n is not even, then n^2 is not even " . Since we are assuming that an integer is either even or odd, this simplifies to " if n is odd, then n^2 is odd " .

Proof of " if n is odd, then n^2 is odd " :
Assume that n is an odd integer, then $n = 2q + 1$ for some $q \in \mathbb{Z}$. Thus,

$$n^2 = (2q + 1)(2q + 1) = 4q^2 + 4q + 1 = 2(2q^2 + 2q) + 1$$

Since $m = 2q^2 + 2q$ is an integer, we have shown that $n^2 = 2m + 1$ and hence n^2 is odd. ■

Another method of proof is the **method of cases**. When this method is used, the original statement is divided into a number of cases that are then proven independently of each other. The cases must be chosen so that they exhaust all possibilities for the hypothesis of the original statement. This method is illustrated in the next example.

Example 3 Prove that if n is an integer, then $n^2 + n$ is even.

Solution We can divide the hypothesis into the following two cases:
Case 1 n is an even integer 　　　　Case 2 n is an odd integer.

Proof in Case 1: Let $n = 2q$ for some $q \in \mathbb{Z}$, then $n^2 + n = (2q)^2 + 2q = 4q^2 + 2q = 2(2q^2 + 1)$. Thus, $n^2 + n$ is even.

Proof in Case 2: Let $n = 2q + 1$ for some $q \in \mathbb{Z}$, then

$$n^2 + n = (2q+1)(2q+1) + (2q+1) = 4q^2 + 6q + 2 = 2(2q^2 + 3q + 1)$$

Thus, $n^2 + n$ is an even integer in this case also. Since we have proved that $n^2 + n$ is even for all possible value of n, then the result is true. ■

Proof By Contradiction

An important method for proving a conditional P → Q is called proof by contradiction. The basic approach is to assume that P is true and Q is false. Then, one reasons to a statement R that contradicts some statement, which is known to be true. This means that the assumption that Q is false is incorrect and hence Q must be true. Therefore P → Q is true. This method of proof is often useful when the conclusion Q is stated in a "negative" form. The next example illustrates this method of proof.

Example 4 For $m,n,p \in \mathbb{Z}$, prove that if $m \mid n$ and $m \nmid p$, then $n \nmid p$.

Proof: Observe that the conclusion $n \nmid p$ has a "negative" form and is difficult to prove directly. Thus, a proof by contradiction will be used. Accordingly, we assume the hypothesis is true and the conclusion is false. Thus, we start with the following.

$$m \mid n \text{ and } m \nmid p \text{ and } n \mid p$$

Since $m \mid n$ and $n \mid p$, there exist integers k and l such that $mk = n$ and $nl = p$. Thus, $(mk)l = p$ and hence $m(kl) = p$. Therefore, $m \mid p$, since $kl \in Z$ because $\mathbb{Z}$ is closed for multiplication. However, we assumed that $m \nmid p$. Therefore, we have reached a contradiction. Consequently $n \mid p$ must be false and therefore $n \nmid p$ must be true. ■

Recall that a rational number has the form $\frac{m}{n}$ where $m,n \in \mathbb{Z}$ and $n \neq 0$. An irrational number can not be written in this form.
The next theorem shows how to prove that certain numbers are irrational.

Theorem 13 The real number $\sqrt{2}$ is irrational.

Proof: We will assume that $\sqrt{2}$ is a rational number and show that a contradiction results. Let $\sqrt{2} = \frac{m}{n}$ for some $m,n \in \mathbb{Z}$ with $n \neq 0$. We can further assume that $\frac{m}{n}$ has been expressed in reduced form so that the only common divisor of m and n is 1. Since $\sqrt{2} = \frac{m}{n}$, we have that $2 = \frac{m^2}{n^2}$ so that $2n^2 = m^2$. Thus, m^2 is an even integer. From Example 2, m is even. Let $m = 2q$ for some $q \in \mathbb{Z}$. Then

$$2n^2 = (2q)^2 = 4q^2$$

Therefore, $n^2 = 2q^2$ and hence n^2 is even. It follows that n is even. Consequently, we have shown that both m and n are even, that is, each has 2 as a divisor. However, this contradicts the original choices for n,m, which were chosen to have no common divisors except 1. Thus, $\sqrt{2}$ can not be a rational number and hence $\sqrt{2}$ is an irrational number. ■

The next theorem provides a method for creating irrational numbers.

Theorem 14 The sum of a rational number and an irrational number is an irrational number.

Proof: Assume that there exists a rational number x and an irrational number y whose sum is a rational number z. Since x and z are rational numbers, we have that $x = \frac{a}{b}$ and $z = \frac{c}{d}$ for some $a, b, c, d \in \mathbb{Z}$ with $b \neq 0$ and $d \neq 0$. Since $x + y = z$ we have

$$y = z - x = \frac{c}{d} - \frac{a}{b} = \frac{bc - ad}{bd}$$

Because $bc - ad$ and bd are integers, the above equation shows that y is a rational number. However, this contradicts the assumption that y is an irrational number. Consequently the sum $x + y$ can not be a rational number. In words, the sum of a rational number and an irrational number must be an irrational number. ■

Summary of Techniques for Proving a Conditional P → Q

We have three primary proof techniques: direct proof, proof by contrapositive, and proof by contradiction. The following chart illustrates these techniques.

	Assume	Show
Direct Proof	P	Q
Contrapositive	~Q	~P
Contradiction	P and ~Q	R, where R contradicts a true statement

The primary difficulty with a proof by contradiction is that we often do not know what statement we are trying to contradict. This means that we must start with P and ~Q and find some resulting statements so that one of these contradicts a statement, which is known to be true.

Sometimes, we must revise the given statement to develop its proof. Consider the following:

If x is odd, then any divisor of x must also be odd.

The conclusion that "any divisor of x must be odd" can be written as

" if $u \mid x$, then u is odd "

Thus, we actually want to prove the following:

If x is odd and $u \mid x$, then u is odd.

In effect, we are using the following logical equivalence:

$$P \to (Q \to R) \approxeq (P \text{ and } Q) \to R$$

Theorem 15 If x is odd and $u \mid x$, then u is odd.

Proof: We use the method of contradiction and assume the hypothesis and negation of the conclusion so that the following is assumed true:

x is odd and $u \mid x$ and u is even

Now, if u is even, then $2 \mid u$. From Corollary 5, it follows that $2 \mid x$, since $u \mid x$ means that x is a multiple of u. Thus, x is even and this makes x an integer that is both even and odd.This contradicts Theorem 12.

The essential idea of this proof is that " if $u \mid x$ and u is even, then x will be even ". Thus, when x is odd, any divisor of x can not be even; otherwise 2 would be a divisor of x and thereby make x even so that a contradiction results. ■

Theorem 16 For $a,b \in \mathbb{R}$, if $ab < 0$, then either $a < 0$ and $b > 0$ or $a > 0$ and $b < 0$.

Proof: The statement

" either $a < 0$ and $b > 0$ or $a > 0$ and $b < 0$ "

means that a and b have "opposite" signs. If this is not true, then a and b have the same signs, that is,

either $a > 0$ and $b > 0$ or $a < 0$ and $b < 0$

Case 1 $a > 0$ and $b > 0$

By Axiom 1 for inequalities, it follows that $ab > 0$, which contradicts the hypothesis that $ab < 0$.

Case 2 $a < 0$ and $b < 0$

Then, $-a > 0$ and $-b > 0$. Thus, $ab = (-a)(-b) > 0$, which again contradicts that $ab < 0$. Consequently, when $ab < 0$, then a and b must have "opposite" signs; otherwise, $ab > 0$ results when a and b have the same sign. ■

Exercises

1. Prove that if $a \mid b$ and $a \mid c$, then $a \mid (b-c)$.

2. Prove that any multiple of an even integer is an even integer.

3. Prove that if $a \mid b$ and c is a multiple of b, then $a \mid c$.

4. Prove that if ab is odd, then a is odd and b is odd.

5. Give a counterexample to " if $a \mid (bc)$ then $a \mid b$ or $a \mid c$ ".

6. Prove that if $a \mid b$ and $c \mid d$, then $(ac) \mid (bd)$.

7. Give a counterexample to " if $a \mid c$ and $b \mid c$, then $(ab) \mid c$ ".

8. Prove that if p and q are primes and $p \mid q$, then $p = q$.

9. Prove that if n^2 is odd, then n is odd.

10. If x is a prime number and $y \in \mathbb{Z}$, what are the possibilities for GCD(x,y) ?

11. Prove for $n \in \mathbb{N}$ and $a \in \mathbb{Z}$, $a \approxeq a(\bmod n)$.

12. Prove that if n is odd, then $n = 4k+1$ for some $k \in \mathbb{Z}$ or $n = 4m-1$ for some $m \in \mathbb{Z}$.

13. Prove that if $d \mid a$ or $d \mid b$, then $d \mid (ab)$.

14. Prove that if $2 \mid a$ and $3 \mid a$, then $6 \mid a$.

15. If $n \in \mathbb{Z}$ and $m = n+1$, then m and n are said to be consecutive integers, Prove or disprove the following:

If m and n are consecutive integers, then $4 \mid (m^2 + n^2 - 1)$.

16. Prove that if $p, q \in \mathbb{Z}$ and $pq = 1$, then $p = 1,\ q = 1$ or $p = -1,\ q = -1$.

17. Prove for $a \neq 0,\ b \neq 0$ that if $a \mid b$ and $b \mid a$, then $a = \pm b$.

18. If x and y are both even, show $4 \mid (x^2 - y^2)$.

19. If x and y are both odd, show $4 \mid (x^2 - y^2)$.

20. If $n \approxeq 0 (\mathrm{mod}\, 3)$ and $n \approxeq 1 (\mathrm{mod}\, 3)$, show $n^2 \approxeq n (\mathrm{mod}\, 3)$.

21. Prove that if $a \approxeq b (\mathrm{mod}\, n)$ and $c \approxeq d (\mathrm{mod}\, n)$, then $(a+c) \approxeq (b+d)(\mathrm{mod}\, n)$.

22. Prove that if $n \mid a$, then $a \approxeq 0 (\mathrm{mod}\, n)$.

23. Prove that if $a \approxeq 5 (\mathrm{mod}\, 6)$, then $a^2 \approxeq 1 (\mathrm{mod}\, 6)$.

24. For $a \in \mathbb{Z}$, prove that if there exist an integer n such that $a \mid (4n+3)$ and $a \mid (2n+1)$, then $a = 1$ or $a = -1$.

25. Prove that n is even if and only if n^3 is even.

26. Prove by contradiction that if a and b are odd, then $4 \nmid (a^2 + b^2)$.

27. Prove that the product of an irrational number and a rational number is irrational.

28. Prove that there is no positive integer x such that $2x < x^2 < 3x$.

29. For $a \in \mathbb{Z}$, prove that $3 \mid a^2$ if and only if $3 \mid a$.
[Hint; every integer can be written as $3q, 3q+1,$ or $3q+2$ for some integer q]

30. Use exercise 29 to prove that $\sqrt{3}$ is irrational.

CHAPTER 4 SETS

Sets and Subsets

A set is a "well-defined" collection of objects. The concept of "well-defined" means that for any given object, it can be determined whether or not the object is a member of the set being investigated. For an example, consider the set of all people in the U.S.A., who have a valid driver's license. A person, who can demonstrate that he or she has a valid driver's license, is then a member of this set; thus, this is a well-defined set. If we examine the set of all tall people in the U.S.A., then this is not a well-defined set because the meaning of "tall" is not specified.

Set Notation

The symbol $\in$ is used to denote set membership. We will generally use upper-case letters to indicate sets and lower-case elements to indicate members of sets. Thus,

$$x \in A$$

stands for " x is a member (or element) of A ". We represent " x is not a member of A " by:

$$x \notin A$$

Definition 1 Two sets A, B are said to be **equal** (written $A = B$), if they have exactly the same elements. We write $A \neq B$ when A and B are not equal.

Definition 2 A set is said to be **finite** provided that it has only a fixed number of elements. Otherwise, the set is called **infinite**.

In **roster notation**, the elements of a set are listed between braces , as in $\{2,4,6,8\}$. The order in which the elements are listed has no significance in determining the set. Thus,

$$\{2,4,6,8\} = \{8,6,4,2\} = \{4,8,6,2\}$$

Infinite sets can also be represented by roster notation. For example,

$$\{1,2,3,\ldots\}$$

indicates the infinite set of all natural numbers or positive integers; here, the " $\ldots$ " stands for "and so on". As before, we use $\mathbb{N}$ to denote the set of all natural numbers.

In **set-builder** notation, let S be the set of all elements under consideration (S is the Universal set) and let $P(x)$ be an open sentence that is either true or false for each element of S , then

$$\{x \in S \mid P(x)\}$$

denotes the set of those elements of S for which $P(x)$ is true. For example, let $S = \mathbb{N}$ and $P(x)$ be the open sentence " $x \leq 5$ " , then

$$\{x \in \mathbb{N} \mid P(x)\} = \{x \in \mathbb{N} \mid x \leq 5\} = \{1,2,3,4,5\}$$

If the set of all integers $\mathbb{Z}$ is the Universal set, then:

$$\{x \in \mathbb{Z} \mid x \leq 5\} = \{\ldots -3,-2,-1,0,1,2,3,4,5\}$$

Definition 3 For A, B any two sets, A **is called a subset of** B (written $A \subseteq B$) provided that every element of the set A is also an element of the set B. If $A \subseteq B$ and $A \neq B$, then A is said to be a **proper subset** of B.

For example,

$$\{1,2,3,4,5\} \subseteq \{\ldots -3,-2,-1,0,1,2,3,4,5\}$$

Example 1 Prove that " if $x \in A$ and $A \subseteq B$, then $x \in B$ ".

Solution Since A is assumed to be a subset of B, every element of A is automatically an element of B. Thus, if x is an element of A, then x must also be an element of B. In symbols: if $x \in A$, then $x \in B$. ■

If the statement $A \subseteq B$ is false, we write $A \nsubseteq B$; this means that there is at least one element of the set A that is not an element of the set B. For example, $A = \{3,4,5\}$ and $B = \{4,5,6,7\}$ satisfy $A \nsubseteq B$.

Definition 4 The set with no elements is called the **empty set** and is denoted by $\varnothing$.

The set of all even integers that are also odd integers is an example of an empty set.

Example 2 Prove that for any set A, the empty set is a subset of A, that is, $\varnothing \subseteq A$.

Solution To prove this assertion, it is helpful to ask what would be required for the statement $\varnothing \subseteq A$ to be false. When $\varnothing \subseteq A$ is false, that is, when $\varnothing \nsubseteq A$ is true, we must be able to find an element of the empty set $\varnothing$ which is not an element of A. Since $\varnothing$ has no elements, then this is impossible. Thus, $\varnothing \subseteq A$ must be true. ■

Many theorems have the biconditional form:

$$S_1 \text{ if and only if } S_2$$

where S_1 and S_2 are statements. Recall that to verify a biconditional, we must prove:

(1) If S_1 then S_2, and

(2) If S_2 then S_1.

In general, to verify an assertion of the form " If S_1 then S_2 ", one assumes that S_1 is true and then shows that the truth of S_2 follows using related definitions and results through a sequence of logical steps. The next theorem and its proof illustrate these methods.

Theorem 1 $A = B$ if and only if $A \subseteq B$ and $B \subseteq A$.

Proof of " If $A = B$ then $A \subseteq B$ and $B \subseteq A$ ".

Suppose that $A = B$. Then, by Definition 1, A and B have exactly the same elements. Therefore, every element of A is also and element of B, that is $A \subseteq B$. Also, every element of B is also an element of A, that is, $B \subseteq A$.

Proof of " if $A \subseteq B$ and $B \subseteq A$, then $A = B$ ".

Now suppose that $A \subseteq B$ and $B \subseteq A$. We must show that $A = B$. The statement $A \subseteq B$ means that every element of A is also an element of B. Moreover, The statement $B \subseteq A$ means that every element of B is also an element of A. Consequently, A and B must have exactly the same elements and hence $A = B$. ■

Definition 5 For any set S, the **power set of** S (written $P(S)$) is defined to be the set of all subsets of S. Thus,

$$P(S) = \{ X \mid X \subseteq S \}$$

The next examples show how to find the power set of S, when S has just one or two elements.

Example 3 Find the power set of the set $\{a\}$ consisting of the single element a.

Solution: We proceed to find all the subsets of $\{a\}$ as follows:

0-element subsets: $\varnothing$

1-element subsets: $\{a\}$

Thus, $P(\{a\}) = \{\varnothing, \{a\}\}$. Note that $P(\{a\})$ has $2^1 = 2$ elements. ■

Example 4 Find the power set of the set $\{a, b\}$ consisting of the two elements a and b.

Solution We proceed to find all the subsets of $\{a, b\}$ as follows:

0-element subsets: $\varnothing$

1-element subsets: $\{a\}, \{b\}$

2-element subsets: $\{a, b\}$

Thus, $P(\{a,b\}) = \{\varnothing, \{a\}, \{b\}, \{a,b\}\}$. Note that $P(\{a,b\})$ has $2^2 = 4$ elements. ■

The observations concerning the number of subsets of sets with one or two elements can be generalized as follows.

Theorem 2 If a set S has n elements, then $P(S)$ has 2^n elements, that is, S has 2^n subsets. Note that we are counting $\varnothing$ and S itself.

The proof of Theorem 2 uses the Principle of Mathematical Induction and is proved later.

Exercises

In Exercises 1-10, let $A = \{1,2,3,4,5,6,7,8,9,10\}$ and $B = \{2,4,6,8,10\}$. Decide whether each statement is TRUE or FALSE and explain your answer.

1. $A \subseteq B$ 2. $B \subseteq A$ 3. $A = B$ 4. $3 \in B$ 5. $3 \notin A$ 6. $3 \in A$ 7. $B \subseteq B$
8. $\varnothing \in A$ 9. $\varnothing \subseteq A$ 10. $5 \in \varnothing$
11. Write in roster notation each of the following sets:

(a) $\{x \in \mathbb{N} \mid x < 7\}$ (b) $\{x \in \mathbb{Z} \mid x < 7\}$

12. Write using set-builder notation each of the following sets:
(a) $\{2,4,6,8,\ldots\}$ (b) $\{7,8,9,10,\ldots\}$

13. List all the subsets of $\{a,b,c\}$ and describe $P(\{a,b,c\})$ in roster notation. [Hint: There are $2^3 = 8$ subsets possible.]

14. For each collection of sets, find the smallest set S such that each of the sets in the collection is a subset of S.
(a) $\{a,b,c\},\{b,d,e\},\{c,d,e,f\}$ (b) $\{2,4,6,8\},\{1,3,5\},\{0\}$

15. Let $A = \{a,b,c,d,e\}$. How many subsets contain the letter e? [Hint: Find how many do not contain e]

16. For sets A,B,C, prove that " if $A \subseteq B$ and $B \subseteq C$, then $A \subseteq C$ ".Observe that this shows that $\subseteq$ has the transitive property.

17. We know that $\varnothing \subseteq A$ for every set A. Is there a set A such that $A \subseteq \varnothing$?

18. Determine $P(\varnothing)$. Does Theorem 2 apply when $n = 0$?

19. Let $P(x)$ be an open sentence that is true or false for elements in the sets A and B. Prove that if $A \subseteq B$, then $\{x \in A \mid P(x)\} \subseteq \{x \in B \mid P(x)\}$.

20. Find $P(\{a,b,c,d\})$.

Unions, Intersections, and Complements

Definition 6 The **union** of two sets A,B (written $A \cup B$) is the set of all elements that belong to A or to B (or possibly belong to both A and B). That is,

$$A \cup B = \{ x \mid x \in A \text{ or } x \in B \}$$

Definition 7 The **intersection** of two sets A,B (written $A \cap B$) is the set of all elements that belong both to A and to B. That is,

$$A \cap B = \{ x \mid x \in A \text{ and } x \in B \}$$

Example 5 Let $A = \{1,2,3,4,5\}$ and $B = \{4,5,6,7\}$, find $A \cup B$ and $A \cap B$.

Solution: $A \cup B = \{1,2,3,4,5,6,7\}$, since these elements are in A or in B. We have that $A \cap B = \{4,5\}$, since these elements are both in A and in B.

Theorem 3 Let A,B be any sets. Then
(a) $A \subseteq A \cup B$ and $B \subseteq A \cup B$ (b) $A \cap B \subseteq A$ and $A \cap B \subseteq B$

Proof of (a):

Suppose that $x \in A$, then $x \in A \cup B$ because $A \cup B = \{x \mid x \in A$ or $x \in B\}$ and hence each element of A is also an element of $A \cup B$. Consequently, A is a subset of $A \cup B$, that is, $A \subseteq A \cup B$. A similar argument shows that $B \subseteq A \cup B$.

Proof of (b):

Suppose that $x \in A \cap B$. By definition of "intersection", this means that $x \in A$ and $x \in B$. In particular, $x \in A$ is true and hence all the elements of $A \cap B$ are also elements of A, that is, $A \cap B \subseteq A$. Similarly, if $x \in A \cap B$, then $x \in B$ follows; thus, $A \cap B \subseteq B$. ■

Definition 8 Sets A and B are called **disjoint**, if $A \cap B = \varnothing$, that is, A and B have no elements in common.

Often we use Venn diagrams to illustrate sets. In a Venn diagram, a set is represented by a circle and the points inside the circle are considered elements of the set, which labels the circle. Regions within more than one set are usually shaded or "dotted".

Example 6 Illustrate $(A \cup B) \cap C$ with a Venn diagram.

Solution: We want to illustrate the set of elements that are in both $A \cup B$ and in C. A preliminary sketch is shown in Figure 1-(a). Here, $A \cup B$ is dotted while C is heavily outlined. By considering the elements that are both in the dotted region for $A \cup B$ and also within the circle for C, we obtain the "dotted" region shown in Figure 1-(b). This represents $(A \cup B) \cap C$.

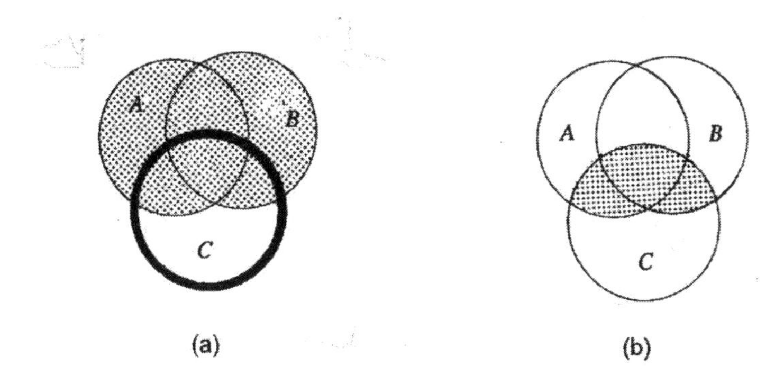

Figure 1

Definition 9 The **universal set** (denoted by U) is the set that contains all sets being discussed or investigated.

In Venn diagrams, U is usually represented as a large rectangle, with all other sets drawn as circles or ovals inside the rectangle.

Definition 10 Let U denote a universal set and let A be a subset of U. The **complement** of a A (written A') is the set of elements in U that are not elements of A. That is,

$$A' = \{x \in U \mid x \notin A\}$$

Figure 2 shows a set A and its complement A', which is shaded.

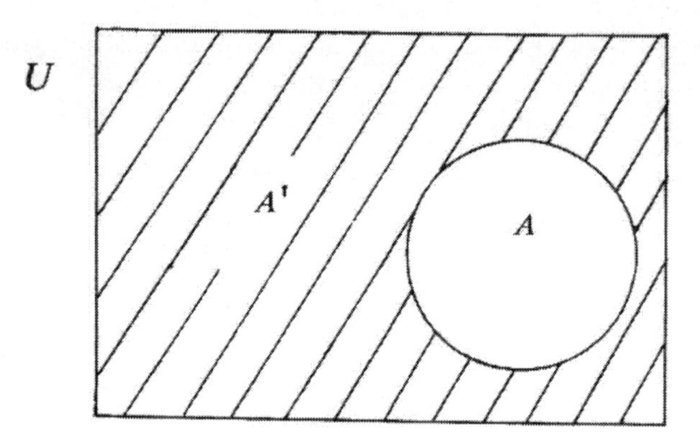

Figure 2

Example 7 Let $U = \{1,2,3,4,5,6,7,8,9,10\}$ and $A = \{1,4,7,10\}$. Find A':

Solution: A' is the set of elements in U that are not in A. Thus, $A' = \{2,3,5,6,8,9\}$.

Theorem 4 Let U be a universal set and let A be any subset of U. Then,
(a) $A \cup A' = U$ (b) $A \cap A' = \varnothing$ (c) $\varnothing' = U$ (d) $U' = \varnothing$ (e) $(A')' = A$

Proof: We prove parts (a) and (b). The other proofs are left as exercises.
To prove (a), we use Theorem 1 and show that both $A \cup A' \subseteq U$ and $U \subseteq A \cup A'$ are true:
First, since U is the universal set, then both $A \subseteq U$ and $A' \subseteq U$ are true. Thus, for any element x in $A \cup A'$, we have $x \in U$ and hence $A \cup A' \subseteq U$.
Now let x be any element of U. Then, either $x \in A$ or $x \notin A$, that is, either $x \in A$ or $x \in A'$. Consequently, $x \in A \cup A'$. Thus, U is a subset of $A \cup A'$, which means that $U \subseteq A \cup A'$.
We have shown that $A \cup A' \subseteq U$ and $U \subseteq A \cup A'$; thus, $A \cup A' = U$ by Theorem 1.

To prove (b), we show that $A \cap A'$ has no elements:
Suppose that $x \in A \cap A'$, then $x \in A$ and $x \in A'$, by definition of "intersection". But $x \in A'$ means that $x \notin A$. Therefore, both the statements $x \in A$ and $x \notin A$ are true, when $x \in A \cap A'$. Since these two statements can not both be true at the same time, it must be the case that $A \cap A'$ has no elements. Thus, $A \cap A' = \varnothing$. ■

Theorem 5 For A, B any sets, $A \subseteq B$ if and only if $A \cap B = A$.

To prove this biconditional statement, we must prove the following two statements:
(i) If $A \subseteq B$, then $A \cap B = A$
(ii) if $A \cap B = A$, then $A \subseteq B$
Proof of " If $A \subseteq B$, then $A \cap B = A$ ":

Suppose that $A \subseteq B$. We must show that $A \cap B = A$, that is, $A \cap B \subseteq A$ and $A \subseteq A \cap B$. From Theorem 3, we know $A \cap B \subseteq A$ is true for any sets A, B. Thus, we need only show that $A \subseteq A \cap B$, when $A \subseteq B$. Let $x \in A$, then it follows that $x \in B$, since $A \subseteq B$. Consequently, we have that $x \in A$ and $x \in B$, that is, $x \in A \cap B$. Thus, $A \subseteq A \cap B$. By Theorem 1, we can conclude that $A = A \cap B$, since we have shown that $A \cap B \subseteq A$ and $A \subseteq A \cap B$.

Proof of " if $A \cap B = A$, then $A \subseteq B$ ":

We now assume that $A \cap B = A$ and show that $A \subseteq B$. Let $x \in A$, then $x \in A \cap B$, since $A \cap B = A$. Thus, $x \in A$ and $x \in B$ are both true. Thus, we have shown that whenever $x \in A$, then $x \in B$ follows. Therefore $A \subseteq B$. ■

Theorem 6 For A, B any sets, $A \subseteq B$ if and only if $A \cup B = B$.

Proof: The proof is similar to the proof of Theorem 5 and is left as an exercise.

We conclude this section by proving DeMorgan's Laws for sets, which simplify finding the complements of unions and intersections.

Theorem 7 DeMorgan's Laws for Sets

For A, B any sets, then (a) $(A \cup B)' = A' \cap B'$ (b) $(A \cap B)' = A' \cup B'$

Proof that $(A \cup B)' = A' \cap B'$:

To use Theorem 1, we must show (i) $(A \cup B)' \subseteq A' \cap B'$ and (ii) $A' \cap B' \subseteq (A \cup B)'$.

To prove (i), we must show that whenever $x \in (A \cup B)'$, then $x \in A' \cap B'$ follows.

Suppose that $x \in (A \cup B)'$, then x is not in $A \cup B$ so that it is false that " $x \in A$ or $x \in B$ ". By DeMorgan's Law of logic, " it is false that $x \in A$ or $x \in B$ " is equivalent to the statement " $x \notin A$ and $x \notin B$ " [~(P or Q) ≊ ~P and ~Q]. By the definition of set complement, $x \notin A$ and $x \notin B$ means that $x \in A'$ and $x \in B'$, that is, $x \in A' \cap B'$. Since we started with $x \in (A \cup B)'$ and showed that $x \in A' \cap B'$, we can conclude that $(A \cup B)' \subseteq A' \cap B'$. This completes the proof of (i).

To prove (ii), we must show that whenever $x \in A' \cap B'$ then $x \in (A \cup B)'$ follows.

Suppose that $x \in A' \cap B'$, then $x \in A'$ and $x \in B'$, that is, $x \notin A$ and $x \notin B$. As in the proof of (i), the statement " $x \notin A$ and $x \notin B$ " is logically equivalent to the statement " it is false that $x \in A$ or $x \in B$ ". Therefore, when $x \in A' \cap B'$, then it is false that $x \in A \cup B$, that is, $x \notin A \cup B$ and consequently $x \in (A \cup B)'$. Since we started with $x \in A' \cap B'$ and showed that $x \in (A \cup B)'$ follows, we can conclude that $A' \cap B' \subseteq (A \cup B)'$. This completes the proof of (ii).

The proof of (b) is left as an exercise. ■

Exercises

1. Let $U = \{1,2,3,4,5,6,7,8,9,10\}$, $A = \{1,2,3,4,5,6\}$, and $B = \{5,6,7,8,9,10\}$. Find:

(a) $A \cup B$ (b) $A \cap B$ (c) A' (d) $A' \cap B'$
(e) $A \cup B'$ (f) $(A \cap B)'$ (g) $(A' \cup B')'$ (h) $(A' \cap B')'$

2. Find an example of three sets such that each pair of these sets is disjoint.

3. Let $A = \{x \in \mathbb{N} \mid 1 \leq x \leq 10\}$, $B = \{x \in \mathbb{N} \mid 5 \leq x \leq 15\}$ and $C = \{2,4,6,8,10,12\}$. Find:
(a) $A \cap B$ (b) $A \cup B$ (c) $A' \cap B$ [Use $U = \mathbb{N}$] (d) $A \cap C$

4. Are the sets $A = \{x \in \mathbb{N} \mid 1 < x < 10\}$ and $B = \{x \in \mathbb{N} \mid 10 < x < 20\}$ disjoint?

In Exercises 5-9, use copies of the Venn diagram shown in Figure 3. Shade in the given regions:

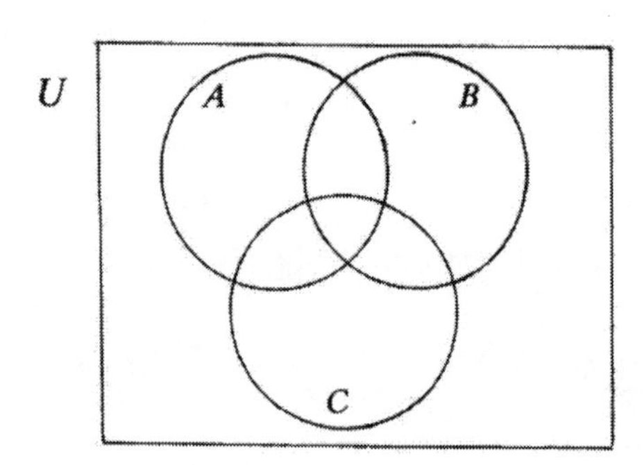

Figure 3

5. $A' \cap B \cup C$ 6. $A' \cup B \cup C$ 7. $A' \cup B'$ 8, $A' \cap B'$ 9. $(A \cup B)' \cap C$

10. For each statement, find specific sets that create a counterexample:

(a) If $A \cap B = A \cap C$, then $B = C$.
(b) If $A \cap B \subseteq C \cap D$, then $A \subseteq C$ and $B \subseteq D$.
(c) If $A \not\subseteq B$ and $B \not\subseteq C$, then $A \not\subseteq C$.

11. Prove that $(A')' = A$.

12, Prove part (b) of Theorem 7, that is, prove that $(A \cap B)' = A' \cup B'$.

In Exercises 13 and 14, use pairs of diagrams like that in Figure 3 to shade in the given sets. For each pair of diagrams created, write an equation that describes the relation between the diagram for (a) and the diagram for '(b).

13. (a) $(A \cap B) \cup (A \cap C)$ (b) $A \cap (B \cup C)$

14. (a) $(A \cup B) \cap (A \cup C)$ (b) $A \cup (B \cap C)$

15. Prove Theorem 6.

16. Prove that if $A \subseteq B$ and $C \subseteq D$, then $A \cup C \subseteq B \cup D$.

17. Prove that if $A \cap B = \varnothing$, then $A \subseteq B'$.

18. Prove that if $A \subseteq B$ and $C \subseteq D$, then $A \cap C \subseteq B \cap D$.

19. Prove that if $A \subseteq B$, then $B' \subseteq A'$.

20. Prove that if $A \not\subseteq B$ and $C \subseteq B$, then $A \not\subseteq C$.

Properties of Set Operations

We now present many of the properties of set operations that we may use in later proofs.

Theorem 8 Let A, B, C be subsets of some universal set U. Then,

(1) Commutative Laws

$A \cap B = B \cap A$ $\qquad\qquad$ $A \cup B = B \cup A$

(2) Associative Laws

$(A \cap B) \cap C = A \cap (B \cap C)$ $\qquad\qquad$ $(A \cup B) \cup C = A \cup (B \cup C)$

(3) Distributive Laws

$A \cap (B \cup C) = (A \cap B) \cup (A \cap C)$ $\qquad\qquad$ $A \cup (B \cap C) = (A \cup B) \cap (A \cup C)$

Proof:

Previously, we have proven that $S = T$ by showing that $S \subseteq T$ and that $T \subseteq S$.
However, we also know that $S = T$ means that " $x \in S$ if and only if $x \in T$, for each x ". We can use this concept to prove that two sets are equal by starting with an element in one set and then using a sequence of " if and only if " statements to show that this is equivalent to being an element of the other set. This approach is used to prove (1) and (2).

Proof that $A \cap B = B \cap A$:

Let $x \in A \cap B$. Then by definition of $A \cap B$,

$$x \in A \cap B \text{ if and only if } x \in A \text{ and } x \in B \qquad \text{(i)}$$

We know that for any statements P,Q that P and Q is logically equivalent to Q and P ; thus,

$$x \in A \text{ and } x \in B \text{ if and only if } x \in B \text{ and } x \in A \qquad \text{(ii)}$$

By definition of $B \cap A$, we have that

$$x \in B \text{ and } x \in A \text{ if and only if } x \in B \cap A \qquad \text{(iii)}$$

Using (i), (ii), and (iii), we conclude that

$$x \in A \cap B \text{ if and only if } x \in B \cap A$$

Therefore, $A \cap B = B \cap A$.
The proof that $A \cup B = B \cup A$ is similar and left as an exercise.

Proof that $(A \cap B) \cap C = A \cap (B \cap C)$:

Let $x \in (A \cap B) \cap C$. Then by definition of $(A \cap B) \cap C$, we have that

$$x \in (A \cap B) \cap C \text{ if and only if } x \in (A \cap B) \text{ and } x \in C \quad \text{(a}$$

From the definition of $(A \cap B)$, we have that

$$x \in (A \cap B) \text{ and } x \in C \text{ if and only if } x \in A \text{ and } x \in B \text{ and } x \in C \quad \text{(b}$$

From the definition of $B \cap C$, we have that

$$x \in A \text{ and } x \in B \text{ and } x \in C \text{ if and only if } x \in A \text{ and } x \in (B \cap C) \quad \text{(c}$$

From the definition of $A \cap (B \cap C)$, we have that

$$x \in A \text{ and } x \in (B \cap C) \text{ if and only if } x \in A \cap (B \cap C) \quad \text{(c}$$

Finally, using (a), (b), (c), and (d), we obtain that

$$x \in (A \cap B) \cap C \text{ if and only if } x \in A \cap (B \cap C)$$

Consequently, $(A \cap B) \cap C = A \cap (B \cap C)$
The proof that $(A \cup B) \cup C = A \cup (B \cup C)$ is similar and left as an exercise.

Proof that $A \cap (B \cup C) = (A \cap B) \cup (A \cap C)$:

We prove that (i) $A \cap (B \cup C) \subseteq (A \cap B) \cup (A \cap C)$ and (ii) $(A \cap B) \cup (A \cap C) \subseteq A \cap (B \cup C)$.

Proof of (i) :

Let $x \in A \cap (B \cup C)$, then $x \in A$ and $x \in (B \cup C)$, that is, $x \in A$ and "$x \in B$ or $x \in C$".
Now, $x \in A$ and $x \in B$ or $x \in C$ is true in the following two cases:

case 1 $x \in A$ and $x \in B$ so that $x \in A \cap B$.

case 2 $x \in A$ and $x \in C$ so that $x \in A \cap C$.

Thus, in either case, we have that $x \in A \cap B$ or $x \in A \cap C$, that is, $x \in (A \cap B) \cup (A \cap C)$.
Since we started with $x \in A \cap (B \cup C)$ and showed that $x \in (A \cap B) \cup (A \cap C)$, we conclude that $A \cap (B \cup C) \subseteq (A \cap B) \cup (A \cap C)$ and (i) is true.

Proof of (ii) :

Let $x \in (A \cap B) \cup (A \cap C)$, then $x \in (A \cap B)$ or $x \in (A \cap C)$.

case 1: $x \in (A \cap B)$, which means that $x \in A$ and $x \in B$.

case 2: $x \in (A \cap C)$, which means that $x \in A$ and $x \in C$.

In either case, we have "$x \in A$ and $x \in B$ or $x \in A$ and $x \in C$" .
This statement is equivalent to "$x \in A$ and $x \in B$ or $x \in C$" , which means that $x \in A$ and $x \in (B \cup C)$ so that $x \in A \cap (B \cup C)$.
Since we started with $x \in (A \cap B) \cup (A \cap C)$ and showed that $x \in A \cap (B \cup C)$, we conclude that $(A \cap B) \cup (A \cap C) \subseteq A \cap (B \cup C)$ and (ii) is true.
The proof that $A \cup (B \cap C) = (A \cup B) \cap (A \cup C)$ is similar and left as an exercise. ■

Sometimes a "direct proof" of a conditional statement P → Q is very difficult. However, since the contrapositive statement ~Q → ~P is equivalent to P → Q , we can prove the contrapositive instead. The next example illustrates this approach.

Theorem 9 Let A,B be subsets of some universal set U. If $A \subseteq B$, then $A \cap B' = \varnothing$.

Proof: To prove that $A \cap B' = \varnothing$ would require proving that $A \cap B'$ has no elements, which is not easily done. Instead we prove the contrapositive " if $A \cap B' \neq \varnothing$, then $A \not\subseteq B$ ".
This requires that we assume that $A \cap B' \neq \varnothing$ and then show that $A \not\subseteq B$ follows.
Suppose that $A \cap B' \neq \varnothing$, then there exist an element x with $x \in A \cap B'$. This means that $x \in A$ and $x \in B'$. Consequently, $x \in A$ and $x \notin B$, by the definition of B'. Thus, we have an element x such that $x \in A$ and $x \notin B$. Thus, $A \not\subseteq B$ since we have found an element, namely, x which is in A but not in B. ■

Cartesian Product of Two Sets

Definition 11 Let A,B be sets. The **Cartesian product**, $A \times B$, of A and B is the set of all ordered pairs (x,y), where $x \in A$ and $y \in B$. We can write the Cartesian product as follows:

$$A \times B = \{ (x,y) \mid x \in A \text{ and } y \in B \}$$

Example 8 Let $A = \{1,2,3\}$ and $B = \{4,5\}$. List the elements of $A \times B$.

Solution: We must find all ordered pairs (x,y) with $x \in A$ and $y \in B$. Thus, we obtain

$$A \times B = \{(1,4),(1,5),(2,4),(2,5),(3,4),(3,5)\}$$ ■

The next theorem states several results for Cartesian products.

Theorem 10 Let A,B,C be sets. Then,

(1) If $A \subseteq B$, then $A \times C \subseteq B \times C$

(2) $A \times (B \cup C) = (A \times B) \cup (A \times C)$

(3) $A \times (B \cap C) = (A \times B) \cap (A \times C)$

Proof of (3) $A \times (B \cap C) = (A \times B) \cap (A \times C)$:
We need to show (i) $A \times (B \cap C) \subseteq (A \times B) \cap (A \times C)$ and (ii) $(A \times B) \cap (A \times C) \subseteq A \times (B \cap C)$.
In proofs with product sets, we must use ordered pairs to represent their elements.
The proof of (i) is given as follows:
Let $(x,y) \in A \times (B \cap C)$, then $x \in A$ and $y \in B \cap C$. Thus, " $x \in A$ and $y \in B$ and $y \in C$ ", which can be written equivalently as " $x \in A$ and $y \in B$ and $x \in A$ and $y \in C$ ". Therefore, $(x,y) \in A \times B$ and $(x,y) \in A \times C$. Consequently, $(x,y) \in (A \times B) \cap (A \times C)$.
Since we started with $(x,y) \in A \times (B \cap C)$ and showed that $(x,y) \in (A \times B) \cap (A \times C)$ follows, we can conclude that $A \times (B \cap C) \subseteq (A \times B) \cap (A \times C)$.
The proof of (ii) is given as follows:
Let $(x,y) \in (A \times B) \cap (A \times C)$, then $(x,y) \in (A \times B)$ *and* $(x,y) \in (A \times C)$.
Thus, $x \in A$ and $y \in B$ *and* $x \in A$ and $y \in C$, which is equivalent to the statement:

" $x \in A$ and $y \in B$ and $y \in C$ ", that is, $x \in A$ and $y \in (B \cap C)$.
This means that $(x,y) \in A \times (B \cap C)$ and hence $(A \times B) \cap (A \times C) \subseteq A \times (B \cap C)$.
The proof of (1) and (2) are similar and left as exercises. ■

Exercises

1. Let $A = \{a,b,c\}$ and $B = \{m,n\}$. List the elements in $A \times B$.

2. In the Cartesian plane, sketch the given regions. Note that $(2,4),(5,7),[1,2],$ and $[3,5]$ are intervals.

(a) $(2,4) \times (5,7)$ (b) $\mathbb{R} \times [1,2]$ (c) $[3,5] \times \mathbb{R}$

3. Give a counterexample to the equality $A \times B = B \times A.$

4. Let A,B be sets. Prove that if $A = B$, then $A \times B = B \times A$.

5. Let A,B,C be sets. Prove that if $A \subseteq B$, then $A \times C \subseteq B \times C$.

6. Let A,B be sets. Prove that $A \times B = \varnothing$ if and only if $A = \varnothing$ or $B = \varnothing$.

7. Let A,B,C be sets. Prove that $A \times (B \cup C) = (A \times B) \cup (A \times C)$.

8. Let A,B,C,D be sets. Prove that if $A \subseteq C$ and $B \subseteq D$, then $A \times B \subseteq C \times D$.

9. Let $A,B,C,$ be sets. Prove that if $A \times B \subseteq A \times C$ and $A \neq \varnothing$, then $B \subseteq C$.

10. Let A,B,C be sets. Prove that $(A \cap B) \times C = (A \times C) \cap (B \times C)$.

CHAPTER 5 PROOFS WITH QUANTIFIERS

Introduction to Quantifiers

In mathematics, many statements use the phrases " for every " or " for each " . One example is

" for every real number x , $x^2 \geq 0$ "

The phrase " for every real number x " is said to quantify the open sentence $(x^2 \geq 0)$ that follows in the sense that it is claiming that $x^2 \geq 0$ is true for all real numbers x .

Definition 1 The phrase " *for every* " (or its equivalents) is called a **universal quantifier**. The symbol $\forall$ is used to denote a universal quantifier.

The statement " for every real number x , $x^2 \geq 0$ " can be written in symbolic form as:

$$(\forall\ x \in \mathbb{R})\ (x^2 \geq 0)$$

From previous results on inequalities, we know that this is a true statement.

An example of a statement using the existential quantifier " *there exists* " is the following:

" There exists an integer x such that $x^2 - 3x - 10 = 0$ "

The phrase " there exists an integer x " is said to quantify the open sentence " $x^2 - 3x - 10 = 0$ " that follows in the sense that it is claiming that the equation " $x^2 - 3x - 10 = 0$ " is true for at least one integer x .

Definition 2 The phrase " *there exists* " (or its equivalents) is called an **existential quantifier**. The symbol $\exists$ is used to denote an existential quantifier

The statement " There exists an integer x such that $x^2 - 3x - 10 = 0$ " has symbolic form:

$$(\exists\ x \in \mathbb{Z})(x^2 - 3x - 10 = 0)$$

Because $x = 5$ makes $x^2 - 3x - 10 = 0$ true, we say that this is a true existential statement.

There are many ways to write statements involving universal quantifiers in English. As an example, we list as follows some equivalent statements for $(\forall\ x \in \mathbb{R})\ (x^2 \geq 0)$.

For any real number x , $x^2 \geq 0$.
The square of every real number is greater than or equal to zero.
If $x \in \mathbb{R}$, then $x^2 \geq 0$.

The example " If $x \in \mathbb{R}$, then $x^2 \geq 0$ " illustrates the fact that conditional statements often contain a "hidden" universal quantifier.

Similarly, There are many ways to write statements involving existential quantifiers in English. As an example, we list as follows some equivalent statements for $(\exists\ x \in \mathbb{Z})(x^2 - 3x - 10 = 0)$:

The equation $x^2 - 3x - 10 = 0$ is true for some integer x.
$x^2 - 3x - 10 = 0$ has an integer solution.
$x^2 - 3x - 10 = 0$ can be solved using an integer x.

Many definitions use universal quantifiers. Consider the following:

Definition 3 For a set $S \subseteq \mathbb{Z}$," S is **closed under addition** " means $(\forall\, x,y \in S)(x + y \in S)$.

Definition 4 For sets A,B , " $A \subseteq B$ " means $(\forall\, x \in A)(x \in B)$.

Also, many definitions use existential quantifiers. This is illustrated as follows:

Definition 5 For $x \in \mathbb{N}$, " x is a **perfect square** " means $(\exists\, q \in \mathbb{N})(q^2 = x)$.

Definition 6 For $x \in \mathbb{N}$, " x is a **composite number** " means there exists a divisor of x other than 1 or x.

The next definition describes the use of the quantifiers " $\forall$ " and " $\exists$ ".

Definition 7 For any open sentence $P(x)$ and set S, then
$(\forall\, x \in S)(\, P(x)\,)$ means every element of S makes $P(x)$ true.
$(\exists\, x \in S)(\, P(x)\,)$ means there is an element of S which makes $P(x)$ true.

Observe that $(\forall\, x \in S)(\, P(x)\,)$ can also be expressed as the conditional statement:

If $x \in S$, then $P(x)$.

For example, $(\forall\, x \in \mathbb{R})\ (x^2 \geq 0)$ can be expressed as " If $x \in \mathbb{R}$, then $x^2 \geq 0$ ".

Negating Quantified Statements

Consider the following statement, which could be true or false depending on the sets A and B:

$$(\forall x \in A)(x \in B) \qquad *$$

This is a formal way of saying that $A \subseteq B$. What does it mean to say * is false?
If we treat * as the conditional " if $x \in A$, then $x \in B$ ", then for this conditional to be false, there must be a counterexample. In other words, there must be an element of A , which is not an element of B. Thus, the negation of * is the existential statement $(\exists x \in A)(x \notin B)$. We write

$$\sim [(\forall x \in A)(x \in B)] \ \approxeq (\exists\, x \in A)(x \notin B)$$

Observe that the statement $(\exists\, x \in A)(x \notin B)$ is a formal statement for $A \not\subseteq B$, which negates the statement $A \subseteq B$. The following theorem generalizes this example.

Theorem 1 $\sim (\forall\, x \in S)(\, P(x)\,)$ is given by $(\exists\, x \in S)(\, \sim P(x)\,)$

Example 1 Find the negation of $(\forall\, a \in \mathbb{N})\,[(\, b \mid a$ and $c \mid a\,) \to (bc) \mid a]$.

Solution Let $P(a)$ be the statement " $[\,(b \mid a$ and $c \mid a\,) \to (bc) \mid a\,]$ "
From Theorem 1, the negation of the original has the form $(\exists\, a \in \mathbb{N})(\sim P(a)\,)$. To find $\sim P(a)$, we must negate the conditional " $[\,(b \mid a$ and $c \mid a\,) \to (bc) \mid a\,]$ " using that $\sim (\,\mathrm{P} \to \mathrm{Q}\,)$ has the form P and ~Q . Therefore, $\sim P(a)$ is given by " $b \mid a$ and $c \mid a$ and $(bc) \nmid a$ ".
Consequently, the negation of the original can be expressed as:

$$(\exists\, a \in \mathbb{N})(\; b \mid a \text{ and } c \mid a \text{ and } (bc) \nmid a\;) \;\blacksquare$$

To show that the statement $(\exists\, a \in \mathbb{N})(\; b \mid a$ and $c \mid a$ and $(bc) \nmid a\;)$ is true, we can use $a = 12,\; b = 4,\; c = 6$ because $4 \mid 12$ and $6 \mid 12$ and $(4)(6) \nmid 12$.
Thus, the original statement $(\forall\, a \in \mathbb{N})[(\, b \mid a$ and $c \mid a\,) \to (bc) \mid a)]$ is false because $a = 12$ is a counterexample. Notice that $a = 12$ is a counterexample only because we could find b and c such that $b \mid 12$ and $c \mid 12$ and $(bc) \nmid 12$.

Consider the following existential statement, which could be true or false depending on the choices for $x, y \in \mathbb{Z}$:

$$(\exists n \in \mathbb{Z})(xn = y) \qquad **$$

In order for ** to be false, the equation $xn = y$ can not be true for any choice of $n \in \mathbb{Z}$ .Thus, $(\forall n \in \mathbb{Z})(xn \neq y)$ and we write

$$\sim (\exists n \in \mathbb{Z})(xn = y) \approxeq (\forall n \in \mathbb{Z})(xn \neq y)$$

The next theorem generalizes this example.

Theorem 2 $\sim (\exists\, x \in S)(\, P(x)\,)$ is given by $(\forall\, x \in S)(\sim P(x)\,)$

Example 2 Find the negation of $(\exists q \in \mathbb{N})(q^2 = n)$

Solution Let $P(n)$ be the statement $q^2 = n$. From Theorem 2, the negation of the original has the form $(\forall q \in \mathbb{N})(\sim P(n))$. This can be written as

$$(\forall q \in \mathbb{N})(q^2 \neq n)$$

Observe that this is a statement about the natural number n and this statement is true when n is not a perfect square. ■

Suppose we want to find a counterexample to the statement

$$(\forall n \in \mathbb{Z})(\cos(n\pi) = -1)$$

A counterexample would be a value of the variable n such that $\cos(n\pi) \neq -1$. The choice $n = 2$ is one obvious counterexample, since $\cos(2\pi) = +1$. Also, $n = 2$ makes the following true:

$$(\exists n \in \mathbb{Z})(\cos(n\pi) \neq -1)$$

In general then, a counterexample to a statement of the form $(\forall\, x \in S)(\, P(x)\,)$ is an element a in S such that $P(a)$ is false. Consequently, a makes its negation $(\exists\, x \in S)(\sim P(x)\,)$ true.

When we negate a statement with more than one quantifier, we consider each quantifier in turn and apply either Theorem 1 or Theorem 2. The next example illustrates this approach.

Example 3 Find the negation of

$$(\exists x \in \mathbb{Z})(\forall y \in \mathbb{Z})(x + y = 0) \qquad *$$

Solution We treat this statement as $(\exists x \in \mathbb{Z})(P(x))$, where $P(x)$ is the open sentence $(\forall y \in \mathbb{Z})(x + y = 0)$. Using Theorem 2, we have that

$$\sim [(\exists x \in \mathbb{Z})(P(x))] \approxeq (\forall x \in \mathbb{Z})(\sim P(x)) \qquad \text{(i)}$$

Next we find $\sim P(x)$. Since $P(x)$ has a universal quantifier, Theorem 1 applies and we have:

$$\sim P(x) \approxeq \sim [(\forall y \in \mathbb{Z})(x + y = 0)] \approxeq (\exists y \in \mathbb{Z})(x + y \neq 0) \qquad \text{(ii)}$$

Combining the results (i) and (ii), we can express the negation of * as:

$$(\forall x \in \mathbb{Z})(\exists y \in \mathbb{Z})(x + y \neq 0) \quad \blacksquare$$

Often we must use results from logic such as DeMorgan's Laws to find negations.

Example 4 Find the negation of

$$(\forall t \in \mathbb{N})[\, t \mid p \rightarrow (t = 1 \text{ or } t = p)] \qquad (*)$$

Solution Let $P(t)$ be the open sentence " $[t \mid p \rightarrow (t = 1 \text{ or } t = p)]$ ". The negation of (*) has the form $(\exists\, t \in \mathbb{N})(\sim P(t))$. Thus, we must find $\sim P(t)$. We use the fact that ~ (P → Q) is given by (P and ~Q) . Hence, $\sim P(t)$ is given by

$$[t \mid p \text{ and } \sim (\, t = 1 \text{ or } t = p)] \approxeq [t \mid p \text{ and } t \neq 1 \text{ and } t \neq p\,]$$

Here, we are using DeMorgan's Law that ~ (P or Q) is logically equivalent to ~P and ~Q , which means that $\sim (\, t = 1 \text{ or } t = p)$ is equivalent to $(t \neq 1 \text{ and } t \neq p)$.
Combining these results, it follows that the negation of (*) can be expressed as

$$(\exists\, t \in \mathbb{N})[\, t \mid p \text{ and } t \neq 1 \text{ and } t \neq p\,]$$

Observe that the original statement (*) is the statement that " p is prime " , whereas its negation given by $(\exists\, t \in \mathbb{N})[\, t \mid p$ and $t \neq 1$ and $t \neq p\,]$ asserts that " p is not prime " . ■

Exercises

1. Write each of the following statements in symbolic form using quantifier notation.

(a) For $S \subseteq \mathbb{Z}$, S is closed under subtraction.

(b) Any set S with n elements has 2^n subsets (S is a subset of some universal set U).

(c) The sum of any three consecutive integers is always divisible by 3 .

(d) x is an integer power of 2 .

(e) The equation " $x^2 = 3$ " has a real solution.

(f) For every real number x , there exists a real number y such that $x + y = 0$.

2. Express the **negation** of each of the following open sentences using quantifier notation.

(a) $(\exists a \in \mathbb{N})(\ b \mid a$ and $c \mid a$ implies $(bc) \mid a\ )$

(b) $(\exists x, y \in \mathbb{Z})(\ ax + by = c\ )$

(c) $(\forall a, b \in \mathbb{R})(\ \sqrt{a+b} \neq \sqrt{a} + \sqrt{b}\ )$

(d) $(\forall m \in \mathbb{Z})(\ m^2 + m$ is odd $)$

(e) $(\forall x \in \mathbb{R})(\exists y \in \mathbb{R})(\ x + y = 0\ )$

(f) $(\forall x, y \in \mathbb{R})(\ x < y$ implies $f(x) < f(y)\ )$

3. For exercise 2 (b), find values for $a, b, c \in \mathbb{Z}$ that make the given statement true and values $a, b, c \in \mathbb{Z}$ that make it false.

4. For exercise 2 (c), find values for $a, b \in \mathbb{R}$ that make the given statement true and values $a, b \in \mathbb{R}$ that make it false.

5. Find all values for $w \in \mathbb{Z}$ that make the statement $(\forall t \in \mathbb{Z})(w \mid t)$ true.

Proofs Involving Quantified Statements

In an existence theorem, the existence of an object possessing some specified property is asserted. A proof of an existence theorem is called an **existence proof**. Such a proof usually requires constructing an object satisfying the specified property. Sometimes, we do not have to find the specific object, but just show that it must exist. For example, the statement " the equation $x^2 - x - 1 = 0$ has a solution x with $1 < x < 2$ " can be proved using the Intermediate Value Theorem without ever finding a specific number x with $1 < x < 2$ that satisfies the equation $x^2 - x - 1 = 0$.

Some examples of existence proofs are now presented.

Example 5 There exist real numbers a and b such that $(a+b)^2 = a^2 + b^2$.

Solution From algebra, we have that $(a+b)^2 = a^2 + 2ab + b^2$. Thus, to make $(a+b)^2 = a^2 + b^2$, we must find numbers a and b with $2ab = 0$. One obvious choice is $a = 1$ and $b = 0$. We then confirm that $(1+0)^2 = 1^2 + 0^2$ is true because $(1+0)^2 = 1^2 = 1$ and $1^2 + 0^2 = 1$ also. ■

Example 6 For any $a, b \in \mathbb{R}$, there exist a real number c with $a < c < b$.

Solution: Viewing a and b on a number line, we expect that the midpoint c between a and b should satisfy $a < c < b$.
Thus, we let $c = \frac{(a+b)}{2}$ and show that $a < c < b$.
Observe that $a < \frac{(a+b)}{2}$ would be true if $2a < a + b$ is true. Since $a < b$, we have that $a + a < a + b$ is true, that is, $2a < a + b$ is true.
Now $\frac{(a+b)}{2} < b$ would be true if $(a + b) < 2b$ is true. Since $a < b$, we have that $a + b < b + b$ is true, that is, $a + b < 2b$ is true.
Thus, we have shown $c = \frac{(a+b)}{2}$ satisfies $a < c$ and $c < b$, that is, $a < c < b$ for $a, b \in \mathbb{R}$. ■

We now discuss in more detail the topic of functions, which uses quantifiers extensively.

Definitions and Proofs About Functions

Definition 8 For sets A and B, f **is a function from** A **to** B (written $f : A \to B$) means for each $u \in A$, f is a method or formula that assigns a unique value $f(u)$ in B. A unique value means " for $u \in A$, if $f(u) = b_1$ and $f(u) = b_2$, then $b_1 = b_2$ ".

Definition 9 For $f : A \to B$, "f is **onto**" (or surjective) means $(\forall b \in B)(\exists a \in A)(f(a) = b)$.

When $f : A \to B$ is onto, then for any $b \in B$, there exists $a \in A$ such that $f(a) = b$.

Example 7 Let $f : \mathbb{R} \to \mathbb{R}$ be defined by $f(x) = 3x - 2$. Show f is onto.

Solution Given any real number y, we must find a real number x with $f(x) = y$.
Thus, we must find x such that $3x - 2 = y$, for any $y \in \mathbb{R}$. By algebra, if $3x - 2 = y$, then $3x = y + 2$ and $x = \frac{(y+2)}{3}$. Since $(y + 2)(\frac{1}{3})$ is a real number, it follows that $x = \frac{(y+2)}{3} = (y + 2)(\frac{1}{3})$ is a real number. Using the rule for the function $f(x)$, we have that

$$f(\frac{(y+2)}{3}) = 3(\frac{(y+2)}{3}) - 2 = (y + 2) - 2 = y$$

This verifies that for any $y \in \mathbb{R}$, the real number $x = \frac{(y+2)}{3}$ satisfies $f(x) = y$. Consequently, f is onto. ■

Definition 10 For a function $f : A \to B$, "f is **one-to-one**" means that $(\forall x_1, x_2 \in A)$(if $f(x_1) = f(x_2)$, then $x_1 = x_2$)

In other words, when f is a one-to-one function, then " if $x_1 \neq x_2$, then $f(x_1) \neq f(x_2)$ ". A one-to-one function is also called an injective function.

Example 8 Let $f(x) = x^2$ define a function from $\mathbb{R}$ to $\mathbb{R}$. Show f is not one-to-one.

Solution We must find $x_1, x_2 \in \mathbb{R}$ such that $x_1 \neq x_2$ and (yet) $f(x_1) = f(x_2)$.
One obvious choice is $x_1 = 2$ and $x_2 = -2$, because then $f(2) = 4 = f(-2)$ and $x_1 \neq x_2$. ■

Theorem 3 Let $f: A \to B$ be a function. Let A have n elements and B have m elements. If $n < m$, then f is not onto.

Proof: We will prove the contrapositive: If f is onto, then $m \leq n$.

Let $b_1, b_2, \ldots, b_m$ be the m distinct elements of B. Then there exist $a_1, a_2, \ldots, a_m$ in A such that $f(a_1) = b_1, f(a_2) = b_2, \ldots, f(a_m) = b_m$, since f is onto.
Now when $i \neq j$, we must have $a_i \neq a_j$ because if $a_i = a_j$ for some $i \neq j$, then $f(a_i) = f(a_j)$, that is, $b_i = b_j$. However, $b_i = b_j$ contradicts the assumption that the $b_1, b_2, \ldots, b_m$ are distinct elements of B. Thus, A has at least the m distinct elements $a_1, a_2, \ldots, a_m$ and hence $m \leq n$, since n is the number of elements in A. This proves that " if f is onto, then $m \leq n$ ". ■

Definition 11 Let $f: \mathbb{R} \to \mathbb{R}$. f **is a strictly increasing function** provided that whenever $x < y$ for $x, y \in \mathbb{R}$ it follows that $f(x) < f(y)$.

The sets $(a,b), (a,b], [a,b), [a,b], (-\infty,a), (-\infty,a], (b,\infty), [b,\infty)$ are examples of intervals in $\mathbb{R}$. All intervals I in $\mathbb{R}$ have the property that whenever $u, v \in I$ and $u < x < v$, then $x \in I$.

Theorem 4 Define $f(I) = \{ y \in \mathbb{R} \mid y = f(x)$, for some $x \in I \}$. Let $f: \mathbb{R} \to \mathbb{R}$ be strictly increasing and onto. Then

If I is an interval in $\mathbb{R}$, then $f(I)$ is an interval in $\mathbb{R}$.

Proof: Let $a, b \in f(I)$, we must show for $y \in \mathbb{R}$ with $a < y < b$, then $y \in f(I)$.
Since f is onto, there exist $u, v, x \in \mathbb{R}$ with $f(u) = a, f(v) = b$, and $f(x) = y$. Now, $u, v \in I$, since $a, b \in f(I)$. If we can show that $u < x < v$, then it follows that $x \in I$, since I is an interval. Consequently, $y = f(x) \in f(I)$ will result and this will prove that $f(I)$ is an interval.
To show $u < x < v$, we will show that $x < u$ and $v < x$ lead to contradictions:
If $x < u$ then $y = f(x) < f(u) = a$ because f is strictly increasing, but this contradicts that $a < y < b$.
If $v < x$ then $b = f(v) < f(x) = y$ because f is strictly increasing, but this also contradicts that $a < y < b$.
Thus, $u < x < v$ must be true. Moreover, since I is an interval with $u, v \in I$, it follows that $x \in I$. Consequently, $f(x) = y \in f(I)$ and hence $f(I)$ is an interval. ■

Definition 12 For functions $f: A \to B$ and $g: B \to C$, the **composition of** f **and** g (written $g \circ f$) is the function defined by $g \circ f(x) = g(f(x))$.

For example if $g(x) = x^2 + 1$ and $f(x) = 3x + 4$, then $g \circ f(x) = (3x+4)^2 + 1$.

Theorem 5 If $f: A \to B$ and $g: B \to C$ are both onto functions, then $g \circ f$ is onto.

Proof: Given any $z \in C$, we must find $x \in A$ such that $g \circ f(x) = z$, that is, $g(f(x)) = z$. Because g is onto, there exists a $y \in B$ such that $g(y) = z$. Moreover, there exists an $x \in A$ such that $f(x) = y$, since f is onto. Consider the following diagram

$$x \xrightarrow{f} y \xrightarrow{g} z$$

We expect that $g \circ f(x) = z$. This is true because $g \circ f(x) = g(f(x)) = g(y) = z$. ■

Disproving Statements

Often we must disprove conditional statements $P \to Q$. If we can write $P \to Q$ in quantified form $(\forall x \in S)(P(x))$, then showing that $(\forall x \in S)(P(x))$ is false is equivalent to showing that $(\exists x \in S)(\sim P(x))$ is true. The next example illustrates this approach.

Example 9 Disprove the following statement:

" if $4 \mid (n^2 - 1)$, then $4 \mid (n-1)$ " *

Solution In quantified form * is $(\forall n \in \mathbb{Z})(4 \mid (n^2 - 1)$ implies $4 \mid (n-1)$). Therefore, the negation of * is

$$(\exists n \in \mathbb{Z})(4 \mid (n^2 - 1) \text{ and } 4 \nmid (n-1))$$

In other words, we must find $n \in \mathbb{Z}$ such that $4 \mid (n^2 - 1$ and $4 \nmid (n-1)$. Keep in mind that our choice for n must make $4 \mid (n^2 - 1)$ true as well as make $4 \nmid (n-1)$ true.
$n = 2$ does not work because $4 \mid (2^2 - 1)$ is not true.
$n = 3$ give $4 \mid (3^2 - 1)$ or $4 \mid 8$ which is true. Also $4 \nmid (n-1)$ is true, since $4 \nmid (3-1)$. Thus, $n = 3$ is a counterexample. ■

Disproving an existential statement is often more difficult than disproving a statement with a universal quantifier.

Example 10 Disprove the statement " there is a real number solution of $x^4 + x^2 + 1 = 0$ ".

Solution: We must prove the negation of $(\exists x \in \mathbb{R})(x^4 + x^2 + 1 = 0)$. Thus, we must prove

$$(\forall x \in \mathbb{R})(x^4 + x^2 + 1 \neq 0)$$

We use the fact that $x^4 \geq 0$ and $x^2 \geq 0$ for every $x \in \mathbb{R}$. Thus, $x^4 + x^2 + 1 \geq 1$, for every $x \in \mathbb{R}$. Consequently, $x^4 + x^2 + 1 \neq 0$, for every $x \in \mathbb{R}$. ■

Example 11 Let $y_1, y_2, y_3, y_4 \in \mathbb{R}$. The mean of these four numbers is computed by

$$y = \frac{(y_1 + y_2 + y_3 + y_4)}{4}$$

Prove that there exist y_i such that $y \leq y_i$.

Solution We expect that the largest of y_1, y_2, y_3, y_4 should work. By relabeling, we can assume that $y_1 \leq y_2 \leq y_3 \leq y_4$.

Observe that $y_1 + y_2 + y_3 + y_4 \leq y_4 + y_4 + y_4 + y_4$. Consequently,

$$y = \frac{(y_1 + y_2 + y_3 + y_4)}{4} \leq \frac{4y_4}{4} = y_4$$

Therefore, $y \leq y_4$. ■

Upper Bounds And The Completeness Axiom

Consider the set $S = (2,7) = \{y \in \mathbb{R} \mid 2 < y < 7\}$. We see that the values $x = 7$ and $x = 8$ make the following statement true:

$$(\forall y \in S)(y \leq x)$$

Therefore, we say that $x = 7$ and $x = 8$ are " upper bounds " for S. The formal definition follows.

Definition 13 For $x \in \mathbb{R}$ and $S \subseteq \mathbb{R}$, x **is an upper bound for** S means $(\forall y \in S)(y \leq x)$.

Observe that sets such as $\mathbb{Z}$ and $S = \{y \in \mathbb{R} \mid 7 < y\}$ do not have any upper bounds. This can be demonstrated for $\mathbb{Z}$ by examining the negation of the statement $(\forall y \in S)(y \leq x)$ which states:

$$(\exists y \in S)(y > x)$$

This statement is true for x when x is *not* an upper bound for S.
Because we know that for any $x \in \mathbb{R}$, there is an integer y such that $x < y$, then x can not be an upper bound for $\mathbb{Z}$ .Thus, $\mathbb{Z}$ has no upper bounds.

If we can find an upper bound for a set $S \subseteq \mathbb{R}$, then we say that S is " bounded above ".

Definition 14 For $S \subseteq \mathbb{R}$, S **is bounded above** means there exist an upper bound x for S, that is,

$$(\exists x \in \mathbb{R})(\forall y \in S)(y \leq x)$$

Theorem 6 Let $S \subseteq \mathbb{R}$ and $x,z \in \mathbb{R}$. If x is an upper bound for S and $x \leq z$, then z is an upper bound for S.

Proof: Assume that x is an upper bound for S and $x \leq z$.
We must show that z is an upper bound for S, that is, $(\forall y \in S)(y \leq z)$.
We use the fact that x is an upper bound for S so that $(\forall y \in S)(y \leq x)$ is true. This means that for any $y \in S$, we have that $y \leq x$. Now, $x \leq z$; thus, $y \leq x \leq z$, for any $y \in S$. Consequently, $(\forall y \in S)(y \leq z)$ and hence z is an upper bound for S. ■

Theorem 7 Let $S,T \subseteq \mathbb{R}$. If S and T are bounded above, then $S \cup T$ is bounded above.

Proof: Suppose that S and T are both bounded above. Let x_S be an upper bound for S and x_T be an upper bound for T. We need to find an upper bound for $S \cup T$. A logical choice is $x = \max\{x_S, x_T\}$, that is, x is the larger of x_S and x_T. We need to show that

$$(\forall y \in S \cup T)(y \leq x)$$ *

Let $y \in S \cup T$, then $y \in S$ or $y \in T$.

Case 1 $y \in S$

Because x_S is an upper bound for S, we have $y \leq x_S$. Therefore, $y \leq x$ because $x_S \leq x$.

Case 2 $y \in T$

Because x_T is an upper bound for T, we have $y \leq x_T$. Therefore, $y \leq x$ because $x_T \leq x$.

In either case, for $y \in S \cup T$, we have that $y \leq x$. Consequently, $(\forall y \in S \cup T)(y \leq x)$ and hence x is an upper bound for $S \cup T$. Therefore, $S \cup T$ is bounded above. ■

Theorem 6 shows that if a set S is bounded above by x, then S will have infinitely many upper bounds; namely, any z with $x \leq z$ is an upper bound for S. We are interested in the " least upper bound ", which is defined as follows:

Definition 15 For $S \subseteq \mathbb{R}$ and $x \in \mathbb{R}$, x **is the least upper bound for** S provided that:

(a) x is an upper bound for S, and

(b) if z is an upper bound for S, then $x \leq z$.

For example, when S is the interval $(3,7) = \{y \in \mathbb{R} \mid 3 < y < 7\}$, then 7 is the least upper bound for S because 7 is an upper bound for S and any number smaller than 7 can not be an upper bound for S. To see this, let $y \in \mathbb{R}$ with $3 < y < 7$, then $x = \frac{(y+7)}{2}$ satisfies $y < x < 7$. Thus, $x \in S$ and $y < x$ so that y can not be an upper bound for S because y is not greater than or equal to every element of S. Thus, 7 is the smallest upper bound for S.

An important property of $\mathbb{R}$ is the following axiom (a statement that we assume without proof).

Completeness Axiom If S is any nonempty subset of $\mathbb{R}$, which has an upper bound, then S has a least upper bound.

The least upper bound for a set that is bounded above may or may not be an element of the set. For example, the interval $S = (3,7)$ has 7 as its least upper bound, but $7 \notin S$. However, the interval $S = (3,7]$ has 7 as its least upper bound and $7 \in S$.

Consider the following set:

$$S = \{3.14,\ 3.141,\ 3.1415,\ 3.14159,\ 3.141592,\ 3.1415926, \ldots\}$$

This set consists of an increasing sequence of numbers that get closer and closer to the irrational number π. Because each number has a finite decimal form, it is a rational number. Moreover, π is the least upper bound for S and $\pi \notin \mathbb{Q}$, since π is an irrational number. Thus, S is a set of rational numbers that is bounded above whose least upper bound is not a rational number. Therefore, the Completeness Axiom does not hold for $\mathbb{Q}$ the set of rational numbers.

The next theorem shows that the least upper bound for a set, which is bounded above, is unique.

Theorem 8 Let $S \subseteq \mathbb{R}$ be bounded above. If l_1 and l_2 are both least upper bounds for S, then $l_1 = l_2$.

Proof: Suppose that l_1 and l_2 are both least upper bounds for S. Then l_1 is an upper bound for

S and l_2 is an upper bound for S. Because l_1 is a least upper bound for S and l_2 is an upper bound, we have $l_1 \leq l_2$.
Similarly, because l_2 is a least upper bound for S and l_1 is an upper bound, we have $l_2 \leq l_1$.
Therefore, $l_1 \leq l_2$ and $l_2 \leq l_1$ are both true; thus, $l_2 = l_1$. ■

Exercises

1. For $a, b \in \mathbb{N}$, prove that if a and b are both perfect squares, then the their product ab is a perfect square.

2. Prove that the sum of any three consecutive integers is divisible by 3.

3. Let $f : A \to B$ and $g : B \to C$ be one-to-one functions. Prove that $g \circ f$ is one-to-one.

4. (a) Use diagrams to show all the possible functions from $\{1,2\}$ to $\{3,4,5\}$.
   (b) How many of these functions are one-to-one?
   (c) How many of these functions are onto?

5. Let $f : \mathbb{R} \to \mathbb{R}$ be defined by $f(x) = 4x + 2$. Show in detail that f is one-to-one and onto.

6. Let $S, T \subseteq \mathbb{R}$ and $S \subset T$. Prove that if T is bounded above, then S is bounded above.

7. Prove or disprove: If $S, T \subseteq \mathbb{R}$ and $S \subset T$ and T is not bounded above, then S is not bounded above.

8. Using the definition of interval, prove or disprove:
(a) If S and T are intervals, then $S \cup T$ is an interval.
(b) If S and T are intervals, then $S \cap T$ is an interval.

9. Prove that if $f : \mathbb{R} \to \mathbb{R}$ is strictly increasing, then f is one-to-one.

10. Let $S, T \subseteq \mathbb{R}$. Suppose that x is an upper bound for both S and T. Prove that x is an upper bound for $S \cup T$.

11. Let $f : A \to B$ and $g : B \to C$. Prove that if $g \circ f$ is onto and g is one-to-one, then f is onto.

12. Prove $(\forall x \in \mathbb{R})(x^2 \leq x$ implies $x \leq 1)$.

13. For $n \in \mathbb{Z}$, prove that if n^4 is even, then $3n + 1$ is odd.

14. Let $f : A \to B$ and let A have n elements and B have m elements. Prove that if f is one-to-one, then $n \leq m$.

15. For $S \subseteq \mathbb{R}$ and $y \in \mathbb{R}$, "y is the maximum element of S" means
$y \in S$ and y is an upper bound for S.

Prove that for $S \subseteq \mathbb{R}$ and $y \in \mathbb{R}$, if y is the maximum element of S, then y is the least upper bound for S.

16. Give an example of two finite sets A and B and a function $f : A \to B$ such that f is one-to-one but not onto.

17. Let $f : \mathbb{R} \to \mathbb{R}$ and $g : \mathbb{R} \to \mathbb{R}$ be functions, define the sum function $f+g$ by $(f+g)(x) = f(x) + g(x)$. Prove that if f and g are both strictly increasing, then $f+g$ is strictly increasing.

18. For $a, b \in \mathbb{Q}$ with $a < b$, prove there is a rational number r such that $a < r < b$.

19. Let A, B, C, D be sets with $A \subseteq B$ and $C \subseteq D$. Prove or disprove that if A and C are disjoint, then B and D are disjoint.

20. Let $f : A \to B$ and $g : B \to C$ with A, B, C nonempty sets. Prove that if $g \circ f$ is one-to-one, then f is one-to-one.

21. Let $a, b, c \in \mathbb{Z}$. Prove that at least one of the integers $a+b,\ a+c,\ b+c$ is even. [Hint: Do proof by cases.]

22. Give an example of a function $f : \mathbb{Z} \to \mathbb{Z}$ that is:
(a) one-to-one and not onto
(b) onto and not one-to-one

23. Let $f : \mathbb{R} \to \mathbb{R}$ be defined by $f(x) = mx + b$, where $m, b \in \mathbb{R}$ and $m \neq 0$. Prove that f is one-to-one.

24. Let $f : \mathbb{R} \to \mathbb{R}$ be defined by $f(x) = x^2 + ax + b$, where $a, b \in \mathbb{R}$. Show that f is not one-to-one. [Hint; consider separately the cases $a \neq 0$ and $a = 0$]

25. Let $f : \mathbb{R} \to \mathbb{R}$ and $g : \mathbb{R} \to \mathbb{R}$ be defined by $f(x) = 7x + 3$ and $g(x) = 3x^2 + 1$. Determine formulas for:
(a) $g \circ f$
(b) $f \circ g$

CHAPTER 6 USING MATHEMATICAL INDUCTION TO PROVE STATEMENTS

We now examine an important method of proof that uses the Principle of Mathematical Induction, which is discussed in this section. A basic set-theoretic concept is given in the following definition.

Definition 1 A set T of numbers is an **inductive set** provided that if $k \in T$, then $k+1 \in T$.

An inductive set is closed under the addition of 1. Intuitively, the natural numbers $\mathbb{N}$ begin with the number 1, after 1 comes 2, after 2 comes 3, and so on. Thus, $\mathbb{N}$ is an inductive set. Other examples of inductive sets are the following.

$$A = \{10, 11, 12, \dots\} = \{n \in \mathbb{N} \mid n \geq 10\}$$

$$B = \{-5, -4, -3, \dots\} = \{n \in Z \mid -5 \leq n\}$$

If T is an inductive set and $1 \in T$, then $1+1 = 2 \in T$, $2+1 = 3 \in T$, and so on. Thus, T should contain all the natural numbers. The essential fact is that

$$\mathbb{N} \text{ is the smallest inductive set that contains } 1. \qquad (*)$$

However, the proof of $(*)$ would be possible only if we had a formal definition of $\mathbb{N}$. We resolve this dilemma by making the statement $(*)$ an axiom for $\mathbb{N}$ so that $(*)$ becomes one of the defining characteristics of the natural numbers.

Axiom 1 (Principle of Mathematical Induction)
Any set which contains the number 1 and which is inductive must contain all the natural numbers.

A more useful form of Axiom 1 is the following version.

Axiom 1′ For any set of numbers T, if $1 \in T$ and T is inductive, then $\mathbb{N} \subseteq T$.

The primary use of Axiom 1 is to prove that a statement $\mathrm{P}(n)$ is true for all natural numbers n. Let T be the truth set for a statement $\mathrm{P}(n)$. If we show that $1 \in T$ and T is inductive, then Axiom $1'$ establishes that $\mathbb{N} \subseteq T$, that is, $\mathrm{P}(n)$ is true for $n \in \mathbb{N}$. Thus, we have the following method.

Method for a Proof by Mathematical Induction
To prove $\mathrm{P}(n)$ is true for each $n \in \mathbb{N}$, the following steps must be completed.

STEP 1 Verify that $\mathrm{P}(1)$ is true. In other words, when $n = 1$ is substituted in the general statement $\mathrm{P}(n)$, the resulting statement should be verifiable.

STEP 2 (Inductive Step) Show that if $\mathrm{P}(k)$ is assumed true, then the statement $\mathrm{P}(k+1)$ can be shown to be true. In doing this, one should first write out both statements $\mathrm{P}(k)$ and $\mathrm{P}(k+1)$. Then, various mathematical results and techniques should be used to derive $\mathrm{P}(k+1)$ from $\mathrm{P}(k)$.

The Inductive Step shows that T (the truth set of $\mathrm{P}(n)$) is an inductive set. The assumption that $\mathrm{P}(k)$ is true is usually called the **inductive hypothesis**.

Example 1 Prove for all natural numbers n that $2^{n-1} \leq n!$

Proof

$\mathrm{P}(1)$ states that $2^0 \leq 1!$. This is true because $2^0 = 1$ and $1! = 1$.

For Step 2, we assume that for $1 \leq k$,

$$2^{k-1} \leq k! \qquad \text{(1-i}$$

We must show that

$$2^{(k+1)-1} \leq (k+1)!$$

In simplified form, we must show that

$$2^k \leq (k+1)! \qquad \text{(1-}$$

We observe that $(k+1)! = (k+1)(k)(k-1)\cdots 1 = (k+1)(k!)$ and $2^k = 2(2^{k-1})$.

Therefore, we multiply both sides of (1-i) by 2 and obtain $2(2^{k-1}) \leq 2(k!)$. By the algebra of exponents, this is equivalent to:

$$2^k \leq 2(k!)$$

If we can show that $2(k!) \leq (k+1)!$, then this will establish (1-ii).

Because $1 \leq k$, we have that $2 \leq k+1$ and therefore $2(k!) \leq (k+1)(k!)$. Thus, we obtain that

$$2^k \leq 2(k!) \leq (k+1)(k!) = (k+1)!$$

This verification that (1-ii) follows from (1-i) shows that $\mathrm{P}(k+1)$ follows from $\mathrm{P}(k)$. Since $\mathrm{P}(1)$ is also true, then $\mathrm{P}(n)$ is true for all natural numbers n . ■

Example 2 Prove if $g(x) = x^n$, then the derivative $g'(x) = n(x^{n-1})$, $n \geq 1$

Proof

For the purposes of this proof, $'$ denotes the operation of taking the derivative.

For $n = 1$, $g(x) = x$ and $g'(x) = 1 = 1x^0$. Thus, $\mathrm{P}(1)$ is true.

For Step 2, assume $\mathrm{P}(k)$ is true; this means that if $k \geq 1$ and $g(x) = x^k$, then

$$g'(x) = (x^k)' = k(x^{k-1}) \qquad ($$

To prove P$(k+1)$, we must prove that if $g(x) = x^{k+1}$, then $g'(x) = (k+1)x^k$.

To show this, we must use statement (2-i) somehow. Since $x^{k+1} = x(x^k)$, then $g(x) = x^{k+1}$ is a product of two functions

$$[p(x) = x][h(x) = x^k]$$

The Product Rule for derivatives states that the derivative of the product function $p(x)h(x)$ is given by the expression $p'(x)h(x) + p(x)h'(x)$. Thus, the derivative of $x^{k+1} = x(x^k)$ is given by

$$(x)'(x^k) + x(x^k)'$$

From (2-i) and $(x)' = 1$, this derivative expression can be written as:

$$(1)(x^k) + x[k(x^{k-1})] \tag{2-ii}$$

By algebra, (2-ii) can be written as:

$$x^k + k[x(x^{k-1})] \;= x^k + k[x^k] = (1+k)x^k = (k+1)x^k$$

Consequently, the derivative of x^{k+1} is given by $(k+1)x^k$. This shows that P$(k+1)$ follows from P(k) . Therefore by the First Principle of Induction, P(n) is true for all natural numbers. ■

Many results involving summation operations can be proved by the techniques of mathematical induction. This is illustrated by the following example.

Example 3 Prove that

$$1 + 2 + \cdots + n = \frac{n(n+1)}{2}$$

Proof

P(1) is clearly true because it states that

$$1 = \frac{1(1+1)}{2} = \frac{2}{2}$$

Now assume that for $k \geq 1$, P(k) is true. This gives the following equation:

$$1 + 2 + \cdots + k = \frac{k(k+1)}{2} \tag{3-i}$$

We must show that P$(k+1)$ is true, that is, we must show that the following equation is true:

$$1 + 2 + \cdots + k + (k+1) = \frac{(k+1)[(k+1)+1]}{2} = \frac{(k+1)(k+2)}{2} \tag{3-ii}$$

We must use equation (3-i) to derive equation (3-ii). We observe that if we were to add $(k+1)$ to the left-hand side of equation (3-i), then we would obtain the left-hand side of equation (3-ii). Therefore, we proceed by adding $(k+1)$ to both sides of equation (3-i) so that the following equation results:

$1 + 2 + \cdots + k + (k+1) = \frac{k(k+1)}{2} + (k+1)$

The right-hand side of the above equation can be manipulated by algebra as follows:

$$\frac{k(k+1)}{2} + (k+1) = \frac{k(k+1)}{2} + \frac{2(k+1)}{2} = \frac{(k+2)(k+1)}{2}$$

This result shows that equation (3-ii) is true. Therefore, P$(k+1)$ follows from P(k) and this proves that P(n) is true for all natural numbers. ■

Example 4 Prove the generalized distributive law for real numbers

$$a(x_1 + x_2 + \cdots + x_n) = ax_1 + ax_2 + \cdots + ax_n \quad \text{(I)}$$

Here n is an "indexing" variable and keeps track of how many real numbers $x_1, x_2, \cdots, x_n$ are being used. We can prove that Formula (I) is true for all natural numbers n by using the basic distributive law, which states that $a(b+c) = ab + ac$.

Proof of Example 4

First, P(1) is clearly true because $a(x_1) = ax_1$.

Next assume P(k) is true, that is,

$$a(x_1 + x_2 + \cdots + x_k) = ax_1 + ax_2 + \cdots + ax_k \quad \text{(4-}$$

We must prove P$(k+1)$ which states that

$$a(x_1 + x_2 + \cdots + x_k + x_{k+1}) = ax_1 + ax_2 + \cdots + ax_k + ax_{k+1} \quad \text{(4}$$

The expression $a(x_1 + x_2 + \cdots + x_k + x_{k+1})$ can be written as $a(b + x_{k+1})$, where $b = (x_1 + x_2 + \cdots + x_k)$. Thus, by the basic distributive law ,

$$a(b + x_{k+1}) = ab + ax_{k+1} = a(x_1 + x_2 + \cdots + x_k) + ax_{k+1}$$

Using (4-i), $a(x_1 + x_2 + \cdots + x_k)$ can be replaced by $ax_1 + ax_2 + \cdots + ax_k$ in the right-hand side of the above equation. Therefore,

$$a(x_1 + x_2 + \cdots + x_k + x_{k+1}) = a(x_1 + x_2 + \cdots + x_k) + ax_{k+1} = ax_1 + ax_2 + \cdots + ax_k + ax_{k+1}$$

This shows that P$(k+1)$ follows from P(k) . Thus, by the First Principle of Induction, (I) is true for all natural numbers n . ■

Example 5 Prove that if $e_1, e_2, \cdots, e_n$ are even integers, then $e_1 + e_2 + \cdots + e_n$ is an even integer.

Proof

If e_1 is even, then the sum with only the term e_1 is equal to e_1 and is therefore an even integer.

Next, we assume that P(k) is true; this means that if $e_1, e_2, \cdots, e_k$ are even integers then the sum $e_1 + e_2 + \cdots + e_k$ is even.

Let $e_1, e_2, \cdots, e_k, e_{k+1}$ be even integers, we must show that the sum $e_1 + e_2 + \cdots + e_k + e_{k+1}$ is even.

The sum $(e_1 + e_2 + \cdots + e_k)$ is even, since we are assuming P(k) is true.

Thus, $(e_1 + e_2 + \cdots + e_k) = 2m$ for some integer m .

Also $e_{k+1} = 2n$ for some integer n , since e_{k+1} is even. Thus, the following is true:

$$e_1 + e_2 + \cdots + e_k + e_{k+1} = (e_1 + e_2 + \cdots + e_k) + e_{k+1} = 2m + 2n = 2(m+n)$$

This shows that the sum $e_1 + e_2 + \cdots + e_k + e_{k+1}$ is a multiple of 2 and therefore an even integer. ■

Example 6 Prove that $5^{2n} - 1$ is a multiple of 8 for all natural numbers n .

Proof:

P(1) is true because when $n = 1$, $5^{2(1)} - 1 = 24 = 8(3)$.

Assume P(k) is true , that is, $5^{2k} - 1 = 8m$ for some integer m . We write this as

$$5^{2k} = 8m + 1 \qquad \text{(6-i)}$$

To show P$(k+1)$ is true, we must show that $5^{2(k+1)} - 1 = 5^{2k+2} - 1$ is a multiple of 8 .

We have the following equation:

$$5^{2k+2} - 1 = 5^{2k}5^2 - 1 = 25(5^{2k}) - 1 \qquad \text{(6-ii)}$$

Now substitute the expression for 5^{2k} from Equation (6-i) into (6-ii) and use algebra to obtain the following chain of equalities:

$$25(5^{2k}) - 1 = 25(8m+1) - 1 = 25(8m) + 25 - 1 = 25(8m) + 24 = 8(25m) + 8(3) = 8(25m+3)$$

This shows that $5^{2(k+1)} - 1$ is a multiple of 8 and that P$(k+1)$ follows from P(k) . ■

Extended Principle of Mathematical Induction

Consider the statement $n^2 < 2^n$. Although this is not true for $n = 2, n = 3,$ and $n = 4$, it appears to be true for $n = 5, 6, 7, \cdots$. To prove this, we need the following extended principle.

Extended Principle of Mathematical Induction

Let M be an integer, If P(n) is a statement such that
(1) P(M) is true;
(2) If P(k) is true, then P$(k+1)$ is true, for $k \geq M$,

then, P(n) is true for all $n \geq M$. That is $\{n \in Z \mid n \geq M\}$ is a subset of the truth set of P(n) .

Example 7 Prove $n^2 < 2^n$, for $n \geq 5$.

Proof

P(5) is true because $5^2 < 2^5$, that is, $25 < 32$ is true.

To show (2) of the Extended Principle, we will assume that $k^2 < 2^k$ is true for $k \geq 5$ and derive

$$(k+1)^2 < 2^{k+1} \qquad \text{(7-i)}$$

In other words, we will assume P(k) is true and show that P($k+1$) follows, when $k \geq 5$.
Multiplying both sides of the inequality $k^2 < 2^k$ by 2 gives

$$2k^2 < 2(2^k) = 2^{k+1} \qquad \text{(7-ii)}$$

If we show that $(k+1)^2 \leq 2k^2$, then (7-i) will follow from (7-ii). We observe that

$$(k+1)^2 = k^2 + 2k + 1$$

Thus, we must show that:

$$k^2 + 2k + 1 \leq 2k^2 = k^2 + k^2$$

This will be true if $2k+1 \leq k^2$. Using the fact that $5 \leq k$, we have that $5k \leq k^2$. Moreover, $2k+1 \leq 5k$, when $k \geq 5$. Therefore,

$$2k + 1 \leq 5k \leq k^2$$

Putting all these results together, we obtain for $k \geq 5$,

$$(k+1)^2 = k^2 + 2k + 1 \leq k^2 + 5k \leq k^2 + k^2 = 2k^2 < 2^{k+1}$$

This shows that P($k+1$) follows from P(k) . ■

Example 8 Prove any natural number n is a sum using only 2's or 3's, if $n \geq 4$.
Note $7 = 2+2+3$ and $17 = 2+2+2+2+3+3+3$.

Proof Because $4 = 2+2$, P(4) is true.

Assume that for $k \geq 4$, P(k) is true, that is, k is a sum using only 2's or 3's. We must show that P($k+1$) is true, that is, $k+1$ is a sum using only 2's or 3's.

Case 1 k can be written as a sum that has at least one 2 .
In this case, $k+1$ would have the term $(2+1) = 3$ as part of its sum. Thus, $k+1$ can be written as a sum using only 2's or 3's .

Case 2 k can **not** be written as a sum that has a least one 2 .
In this case, k is equal to a sum that uses only 3's . Consequently, $k+1$ would have the term $(3+1) = 2+2$ as part of its sum. Thus, $k+1$ can be written as a sum using only 2's or 3's .

By the Extended Principle of Induction, P(n) is true for $n \geq 4$. ■

Example 9 Prove that $(\cos x + i\sin x)^n = \cos(nx) + i\sin(nx)$ for $n \geq 1$.

Here $i^2 = -1$. We will use the following trigonometric identities for $\cos(a+b)$ and $\sin(a+b$

$$\cos(a+b) = \cos a\cos b - \sin a\sin b \text{ and } \sin(a+b) = \sin a\cos b + \cos a\sin b$$

Proof

When $n = 1$, then P(1) states that $(\cos x + i\sin x)^1 = \cos(1(x)) + i\sin(1(x))$ which is clearly true.
Assume for $k \geq 1$, that

$$(\cos x + i\sin x)^k = \cos(kx) + i\sin(kx) \qquad \text{(9-i)}$$

We must prove $P(k+1)$, that is:

$$(\cos x + i\sin x)^{k+1} = \cos[(k+1)x] + i\sin[(k+1)x] \qquad \text{(9-ii)}$$

From the rules of exponents, we know that $(\cos x + i\sin x)^{k+1} = (\cos x + i\sin x)(\cos x + i\sin x)^k$
From equation (9-ii), this becomes
$(\cos x + i\sin x)^{k+1} = (\cos x + i\sin x)[\cos(kx) + i\sin(kx)]$

Multiplying terms, the right-hand side of above equation is then given by

$$\cos x\cos(kx) + i\cos x\sin(kx) + i\sin x\cos(kx) + i^2\sin x\sin(kx)$$

Collecting terms and using $i^2 = -1$, this can be written as follows

$$[\cos x\cos(kx) - \sin x\sin(kx)] + i\,[\sin x\cos(kx) + \cos x\sin(kx)] \qquad \text{(9-iv)}$$

If we let $a = x$ and $b = kx$ in the identities for $\cos(a+b)$ and $\sin(a+b)$, then (9-iv) becomes

$$\cos(x+kx) + i\sin(x+kx) = \cos[(1+k)x] + i\sin[(1+k)x]$$

Therefore, we have shown that if $P(k)$ is true, then

$$(\cos x + i\sin x)^{k+1} = \cos[(1+k)x] + i\sin[(1+k)x]$$

In other words, $P(k+1)$ follows from $P(k)$.
Consequently, by the First Principle of Induction, $P(n)$ is true for all natural numbers n . ■

Example 10 Prove that if $n \geq 1$ and a, b are real numbers then

$$\begin{bmatrix} a & 0 \\ 0 & b \end{bmatrix}^n = \begin{bmatrix} a^n & 0 \\ 0 & b^n \end{bmatrix}$$

Proof

P(1) is true because when $n = 1$, $P(n)$ states that

$$\begin{bmatrix} a & 0 \\ 0 & b \end{bmatrix}^1 = \begin{bmatrix} a^1 & 0 \\ 0 & b^1 \end{bmatrix}$$

For $k \geq 1$, assume $P(k)$ is true, that is, the following is true:

$$\begin{bmatrix} a & 0 \\ 0 & b \end{bmatrix}^k = \begin{bmatrix} a^k & 0 \\ 0 & b^k \end{bmatrix}$$

We must prove that:

$$\begin{bmatrix} a & 0 \\ 0 & b \end{bmatrix}^{k+1} = \begin{bmatrix} a^{k+1} & 0 \\ 0 & b^{k+1} \end{bmatrix}$$

By exponent rules,

$$\begin{bmatrix} a & 0 \\ 0 & b \end{bmatrix}^{k+1} = \begin{bmatrix} a & 0 \\ 0 & b \end{bmatrix}\begin{bmatrix} a & 0 \\ 0 & b \end{bmatrix}^k$$

From the assumption that P(k) is true, the right-hand side of the above equation is given by

$$A = \begin{bmatrix} a & 0 \\ 0 & b \end{bmatrix}\begin{bmatrix} a^k & 0 \\ 0 & b^k \end{bmatrix}$$

Denote the resulting product matrix A by

$$\begin{bmatrix} a_{11} & a_{12} \\ a_{21} & a_{22} \end{bmatrix}$$

Using matrix multiplication, we can compute the entries of A by the following "dot" products:

$$a_{11} = \begin{bmatrix} a & 0 \end{bmatrix}\begin{bmatrix} a^k \\ 0 \end{bmatrix} = a(a^k) + 0(0) = a^{k+1}$$

$$a_{12} = \begin{bmatrix} a & 0 \end{bmatrix}\begin{bmatrix} 0 \\ b^k \end{bmatrix} = a(0) + 0(b^k) = 0$$

$$a_{21} = \begin{bmatrix} 0 & b \end{bmatrix}\begin{bmatrix} a^k \\ 0 \end{bmatrix} = 0(a^k) + b(0) = 0$$

$$a_{22} = \begin{bmatrix} 0 & b \end{bmatrix}\begin{bmatrix} 0 \\ b^k \end{bmatrix} = 0(0) + b(b^k) = b^{k+1}$$

Therefore,

$$\begin{bmatrix} a & 0 \\ 0 & b \end{bmatrix}^{k+1} = \begin{bmatrix} a_{11} & a_{12} \\ a_{21} & a_{22} \end{bmatrix} = \begin{bmatrix} a^{k+1} & 0 \\ 0 & b^{k+1} \end{bmatrix}$$

This shows that P$(k+1)$ follows from P(k). Thus, by the First Principle of Induction P(n) is true for all n. ■

For a set A, we represent the complement of A by A'. From DeMorgan's Laws, we have that $(A \cup B)' = A' \cap B'$. The next example generalizes this result to n sets.

Example 11 For $A_1, A_2, \cdots, A_n$ any n sets, prove that

$$(A_1 \cup A_2 \cup \cdots \cup A_n)' = A_1' \cap A_2' \cap \cdots \cap A_n'$$

Proof

For $n = 1$, the equation $A_1' = (A_1)'$ is obviously true.

Assume for $k \geq 1$ that

$$(A_1 \cup A_2 \cup \cdots \cup A_k)' = A_1' \cap A_2' \cap \cdots \cap A_k' \qquad (11\text{-i})$$

We must use this to prove $\mathrm{P}(k+1)$, that is, we must derive the following equation

$$(A_1 \cup A_2 \cup \cdots \cup A_k \cup A_{k+1})' = A_1' \cap A_2' \cap \cdots \cap A_k' \cap A_{k+1}' \qquad (11\text{-ii})$$

Let $L = A_1 \cup A_2 \cup \cdots \cup A_k$, then by DeMorgan's Laws

$$(A_1 \cup A_2 \cup \cdots \cup A_k \cup A_{k+1})' = (L \cup A_{k+1})' = L' \cap A_{k+1}' \qquad (11\text{-iii})$$

From (11-i), $L' = A_1' \cap A_2' \cap \cdots \cap A_k'$. Thus the right-hand side of (11-iii) satisfies

$$L' \cap A_{k+1}' = (A_1' \cap A_2' \cap \cdots \cap A_k') \cap A_{k+1}' = A_1' \cap A_2' \cap \cdots \cap A_k' \cap A_{k+1}'$$

This shows that (11-ii) is true and thereby verifies that $\mathrm{P}(k+1)$ follows from $\mathrm{P}(k)$. Thus, by the First Principle of Induction, $\mathrm{P}(n)$ is true for all natural numbers n. ■

Exercises

1. Prove by induction that for each natural number n, each of the following is true.

(a) $1^2 + 2^2 + \cdots + n^2 = \frac{(n+1)(2n+1)}{6}$

(b) $2 + 4 + 6 + \cdots + 2n = n(n+1)$

(c) $3 + 6 + 9 + \cdots + 3n = \frac{3n(n+1)}{2}$

(d) $\frac{1}{1(2)} + \frac{1}{2(3)} + \frac{1}{3(4)} + \cdots + \frac{1}{n(n+1)} = \frac{n}{(n+1)}$

(e) $1 + 2 + 2^2 + 2^3 + \cdots + 2^n = 2^{(n+1)} - 1$

(f) $1 + a + a^2 + a^3 + \cdots + a^n = \frac{a^{n+1}-1}{a-1}$

(g) $5^n - 2^n$ is a multiple of 3

(h) $3^n - 1$ is a multiple of 2

(i) $n^3 - n$ is a multiple of 3

(j) $9^n - 4^n$ is a multiple of 5

2. Use $\cos(a+b) = \cos a \cos b - \sin a \sin b$ and $\cos(\pi) = -1$ to prove $\cos(n\pi) = (-1)^n$

3. Use the Extended Principle of Induction to prove each of the following for the subset of natural numbers indicated:

(a) $n^3 < 2^n$ for $n \geq 10$

(b) $2^{n+1} \leq 3^n$ for $n \geq 2$

(c) $4^n < n!$ for $n \geq 9$

(d) $n^4 < 2^n$ for $n \geq 17$

(e) $n^2 \leq n!$ for $n \geq 4$

4. Prove each of the following for all natural numbers.

(a) If $1 < x$, then $1 < x^n$

(b) If $0 < a < b$, then $0 < a^n < b^n$

(c) If $x \geq 1$, then $x^n \geq x$

5. If $x \mid a_1$, $x \mid a_2$,..., $x \mid a_n$, then $x \mid (a_1 + a_2 + \cdots + a_n)$.

6. Use $|x+y| \leq |x| + |y|$ to prove $|x_1 + x_2 + \cdots + x_n| \leq |x_1| + |x_2| + \cdots + |x_n|$ for real numbers $x_1, x_2, \cdots, x_n$.

7. Use $(A \cap B)' = A' \cup B'$ to prove that if $A_1, A_2, \ldots, A_n$ are any n sets, then

$$(A_1 \cap A_2 \cap \cdots \cap A_n)' = A_1' \cup A_2' \cup \cdots \cup A_n'$$

8. A basic fact of algebra states that " if $xy = 0$, then $x = 0$ or $y = 0$, for real numbers x, y " . Use this fact and induction to prove that if $x_1, x_2, \ldots, x_n$ are real numbers such that $(x_1)(x_2)\cdots(x_n) = 0$, then at least one of the numbers $x_1, x_2, \ldots, x_n$ is 0 .

9. Prove that if $x_1, x_2, \ldots, x_n$ are odd integers , then their product $(x_1)(x_2)\cdots(x_n)$ is odd.

10. Prove every natural number with $n \geq 7$ can be written as a sum using only 2's and 5's .

11. Suppose a real-valued function $g(x)$ having domain all real numbers has the property $g(x+T) = g(x)$. Prove that $g(x+nT) = g(x)$ for all real numbers x .

12. Prove $\begin{bmatrix} 1 & 0 \\ 0 & 1 \end{bmatrix} \cdot \begin{bmatrix} 2 & 0 \\ 0 & 2 \end{bmatrix} \cdots \cdot \begin{bmatrix} n & 0 \\ 0 & n \end{bmatrix} = \begin{bmatrix} n! & 0 \\ 0 & n! \end{bmatrix}$

13. A basic property of the rational number system Q is closure under addition, that is, if $a, b \in Q$ then $a+b \in Q$. Use this fact and induction to prove that if $x_1, x_2, \ldots, x_n$ are n rational numbers then their sum $x_1 + x_2 + \cdots + x_n$ is a rational number.

14. For sets A, B, C , we have $A \cap (B \cup C) = (A \cap B) \cup (A \cap C)$. Prove for any sets $A, B_1, B_2, \ldots, B_n$ that

$$A \cap (B_1 \cup B_2 \cup \cdots \cup B_n) = (A \cap B_1) \cup (A \cap B_2) \cup \cdots \cup (A \cap B_n)$$

15. Let $g(x) = xe^{-x}$. Prove that the nth derivative $g^n(x)$ is given by $g^n(x) = (-1)^n e^{-x}(x-n)$.

The Second Principle of Mathematical Induction

An important result is the following theorem.

Theorem 1 For every natural number $n \geq 2$, either n is prime or n is a product of primes.

If we examine numbers such as $k = 55 = 11(5)$ and $k+1 = 56 = 2(2)(2)(7)$, we see that there are no techniques for using the prime factors of k to obtain the prime factors of $k+1$. In other words, if Theorem 1 is true for $n = k$, then we can not use this to prove that Theorem 1 is true for $n = k+1$.

To prove Theorem 1, we will need the Second Principle of Induction, which can be formulated as follows:

Second Principle of Mathematical Induction

Let M be an integer. If P(n) is a statement such that

1. P(M) is true for $M \in Z$, and
2. For every $k \geq M$, if P(M) , P$(M+1)$,... ,P(k) are true, then P$(k+1)$ is true,

then P(n) is true for all integers $n \geq M$.

To use the Second Principle to prove that a statement P(n) is true for all integers $n \geq M$ requires the completion of the following two steps:

STEP A Verify that P(M) is true.

STEP B Verify that if P(M) , P($M+1$) ,..., P(k) are assumed true for $k \geq M$, then the statement P($k+1$) can be shown to be true.

We illustrate this method in the next example.

Example 12 Prove Theorem 1.

Proof

STEP A P(2) is directly true, since 2 is a prime number.

STEP B Assume that for $k \geq 2$, P(2) , P(3) ,⋯, P(k) are true, that is, we assume that each of the integers $2,3,\ldots,k$ is a prime number or a product of prime numbers. We must prove that $k+1$ is a prime number or product of prime numbers.

Case 1 $k+1$ is a prime number.
In this case, P($k+1$) is directly true.

Case 2 $k+1$ is **not** a prime number.

In this case, $k+1$ can be written as a product of two natural numbers x,y such that $2 \leq x \leq k$ and $2 \leq y \leq k$. In other words, since $k+1$ is not prime, it has factors x,y other than 1 and itself.

Because $2 \leq x \leq k$ and $2 \leq y \leq k$, then P(x) and P(y) are both true . Thus, x is prime or a product of prime factors and y is prime or a product of prime factors. There are four possibilities.

(1) x is prime and y is prime. Thus, $k+1 = xy$ is a product of primes.

(2) x is prime and $y = q_1q_2\cdots q_n$ is a product of primes $q_1,q_2,\ldots,q_n$. Thus, $k+1 = xy = x(q_1q_2\cdots q_n)$ is a product of primes.

(3) y is prime and $x = p_1p_2\cdots p_m$ is a product of primes $p_1,p_2,\ldots,p_m$. Thus, $k+1 = xy = (p_1p_2\cdots p_m)y$ is a product of primes.

(4) $x = p_1p_2\cdots p_m$ is a product of primes $p_1,p_2,\ldots,p_m$ and $y = q_1q_2\cdots q_n$ is a product of primes $q_1,q_2,\ldots,q_n$. Thus, $k+1 = xy = (p_1p_2\cdots p_m)(q_1q_2\cdots q_n)$ is a product of primes.

Thus, we have shown that P($k+1$) follows, when we assume that P(2) , P(3) ,⋯, P(k) are true.

By the Second Principle of Induction, this establishes that P(n) is true for $n \geq 2$. ■

As the previous example illustrates, the Second Principle is used when the proof of the statement P($k+1$) during the Inductive Step is not based solely on the assumption that P(k) is true, but requires that **all** the statements P(1), P(2),..., P(k) be assumed true.

To analyze set-theoretic concepts that are used to make the connection between the First Principle and the Second Principle, we make the following definition.

Definition 2 For $S \subseteq \mathbb{N}$, S is **weakly inductive** if $\{1,2,\ldots,k\} \subseteq S$ implies $k+1 \in S$.

Theorem 2 (Set-theoretic version of the Second Principle of Induction)

Any subset of $\mathbb{N}$ which contains the number 1 and which is weakly inductive must be equal to $\mathbb{N}$.

Theorem 2 can be proved using the First Principle of Induction as follows:

Proof of Theorem 2

Assume that $S \subseteq \mathbb{N}$ and $1 \in S$ and S is weakly inductive. We must prove that $S = \mathbb{N}$.

Let P(n) be the statement " $\{1,2,\ldots,n\} \subseteq S$ ". We will prove that P(n) is true for all $n \in \mathbb{N}$. Because $n \in \{1,2,\ldots,n\}$, this will show that $n \in S$ for all $n \in \mathbb{N}$ as desired.

First P(1) is true since $\{1\} \subseteq S$ is true from the assumption that $1 \in S$.

Next, assume that P(k) is true, that is, $\{1,2,\ldots,k\} \subseteq S$, then it follows that $k+1 \in S$ because S is weakly inductive. Therefore,

$$\{1,2,\ldots,k\} \cup \{k+1\} = \{1,2,\ldots,k+1\} \subseteq S$$

Thus, P($k+1$) follows from P(k). Consequently by Axiom $1'$, the statement $\{1,2,\ldots,n\} \subseteq S$ is true for all $n \in \mathbb{N}$ and therefore $S = \mathbb{N}$. ■

Theorem 2 shows that the First Principle can be used to prove the Second Principle. Conversely, the Second Principle can be used to prove the First Principle; this is left as an exercise.

Exercises

1. Use the Second Principle of Induction to prove directly the following statement:
   "If n is a natural number greater than 1, then n is divisible by some prime."

2. Prove that any inductive set is weakly inductive.

3. Use Exercise 2 to prove that the First Principle of Induction follows from the Second Principle.

4. Recall that a set of numbers S has a least element means $\exists\ x_0 \in S$ such that $x_0 \leq x$ for each $x \in S$.
Without using the Second Principle, prove that conditions (a) and (b) are equivalent. Assume $S \subseteq \mathbb{N}$.

(a) $1 \in S$ and S is weakly inductive.

(b) $\mathbb{N} - S$ has no least element.
[Hint: to prove $(a) \Rightarrow (b)$, use proof by contradiction]

CHAPTER 7 RELATIONS AND FUNCTIONS

Relations – Definitions and Examples

Recall that a function $f : A \to B$ can be thought of as a set of ordered pairs that is a special type of subset of $A \times B$. Actually functions are a special case of a relation from the set A to the set B.

Definition 1 A **relation** R from the set A to the set B is any subset of $A \times B$.

There are many familiar examples of mathematical relations; some of these are the following:

Example 1 (some mathematical relations)

Description as a set of ordered pairs	Open sentence
1. $R = \{(x,y) \in \mathbb{R} \times \mathbb{R} \mid x \le y\}$	$x \le y$
2. $R = \{(x,y) \in \mathbb{R} \times \mathbb{R} \mid \mid x \mid = \mid y \mid\}$	$\mid x \mid = \mid y \mid$
3. $R = \{(m,n) \in \mathbb{Z} \times \mathbb{Z} \mid m \mid n\}$	$m \mid n$
4. $R = \{(X,Y) \in P(S) \times P(S) \mid X \subseteq Y\}$	$X \subseteq Y$

We sometimes write $x \sim y$ to denote that x is related to y by a specified condition. For example, for $(m,n) \in \mathbb{N} \times \mathbb{N}$, $m \sim n$ if and only if m and n have the same prime divisors.

Example 2 (some nonmathematical relations)

1. $A =$ the set of states in the U.S.A.
   $R = \{(x,y) \in A \times A \mid x$ and y have a border in common$\}$

2. $A =$ the set of countries in the world.
   $R = \{(x,y) \in A \times A \mid x$ has a smaller population than $y\}$

The concepts of domain and range of a function can be extended to relations as follows.

Definition 2 For R any relation from the set A to the set B :

The **domain** of R [written $Dom(R)$] is

$$\{x \in A \mid \exists\, y \in B \textit{ with } (x,y) \in R\}$$

The **image** of R or range of R [written $\mathrm{Im}(R)$] is

$$\{y \in B \mid \exists\, x \in A \textit{ with } (x,y) \in R\}$$

Example 3 Let $R_1 = \{(x,y) \in \mathbb{R} \times \mathbb{R} \mid y = \sin x\}$ and $R_2 = \{(x,y) \in \mathbb{R} \times \mathbb{R} \mid y = \tan x\}$

Find the domain and image of R_1 and R_2 .

Solution

(a) The domain of R_1 is the set of all real numbers x for which $\sin x$ is defined, that is $Dom(R_1) = \mathbb{R}$.
The image of R_1 is the set of all y values that equal $\sin x$ for some real number x ; thus, $\text{Im}(R_1) = \{y \in \mathbb{R} \mid -1 \leq y \leq 1\}$.

(b) The domain of R_2 is the set of all real numbers x for which $\tan x$ is defined; thus, $Dom(R_2) = \{x \in \mathbb{R} \mid x \neq n(\frac{\pi}{2}), n = \pm 1, \pm 3, \ldots\}$.
The image of R_2 is the set of all y values that equal $\tan x$ for some real number x ; thus, $\text{Im}(R_1) = \mathbb{R}$. ■

Inverse Relations

Given any relation R from the set A to the set B , we define the inverse relation R^{-1} from the set B to the set A as follows.

Definition 3 Let R be a relation from the set A to the set B . The **inverse of** R , written R^{-1} , is the relation from B to A defined by

$$R^{-1} = \{(y,x) \in B \times A \mid (x,y) \in R\}$$

That is, R^{-1} is the subset of $B \times A$ consisting of all ordered pairs (y,x) such that $(x,y) \in R$.

Example 4

(a) If $R = \{(1,2),(3,4),(5,6)\}$ then $R^{-1} = \{(2,1),(4,3),(6,5)\}$

(b) If $R = \{(x,y) \in \mathbb{R} \times \mathbb{R} \mid y = 3x\}$ then $R^{-1} = \{(a,b) \in \mathbb{R} \times \mathbb{R} \mid b = \frac{a}{3}\}$

(c) if $D = \{(m,n) \in \mathbb{Z} \times \mathbb{Z} \mid m \mid n\}$ then

$D^{-1} = \{(n,m) \in \mathbb{Z} \times \mathbb{Z} \mid m \mid n\} = \{(n,m) \in \mathbb{Z} \times \mathbb{Z} \mid n \text{ is a multiple of } m\}$

Theorem 1 If R is a relation from the set A to the set B then

(1) $Dom(R^{-1}) = \text{Im}(R)$
(2) $\text{Im}(R^{-1}) = Dom(R)$
(3) $(R^{-1})^{-1} = R$

Proof of (1)

We need to prove $Dom(R^{-1}) \subseteq \text{Im}(R)$ and $\text{Im}(R) \subseteq Dom(R^{-1})$.

To prove that $Dom(R^{-1}) \subseteq \text{Im}(R)$, we let $y \in Dom(R^{-1})$ and then show that $y \in \text{Im}(R)$. By definition of $Dom(R^{-1})$ and the fact that R^{-1} is a relation from B to A, there exists an $x \in A$ such that $(y,x) \in R^{-1}$. Consequently, $(x,y) \in R$, by definition of R^{-1}. Therefore, $y \in \text{Im}(R)$, by definition of image.

To prove that $\text{Im}(R) \subseteq Dom(R^{-1})$, we let $y \in \text{Im}(R)$ and show that $y \in Dom(R^{-1})$. By definition of $\text{Im}(R)$ and the fact that R is a relation from A to B, there exists an $x \in A$ such that $(x,y) \in R$ and therefore $(y,x) \in R^{-1}$. Consequently, $y \in Dom(R^{-1})$, by definition of domain of a relation.

The proofs of (2) and (3) of Theorem 1 are left as exercises. ■

Invertible Functions

An important result for functions that follows from Theorem 1 is the following.

Corollary 2 For any function $f : A \to B$ we have $Dom(f^{-1}) = B$ if and only if f is onto.

Proof

Recall that $f : A \to B$ is onto if and only if for every $b \in B$, there exist an $a \in A$ such that $f(a) = b$. This means that $f : A \to B$ is onto if and only if $\text{Im}(f) = B$. By Theorem 1, $Dom(R^{-1}) = \text{Im}(R)$ for relations and in particular for functions. Thus, $Dom(f^{-1}) = B$ if and only if $\text{Im}(f) = B$ if and only if f is onto. ■

If $f : A \to B$ is a function, then f^{-1} is a relation from B to A. What properties must f have in order that f^{-1} be a function from B to A? From our previous discussion of functions, we know that if f^{-1} is a function, then f^{-1} must have the following two properties.

(i) $Dom(f^{-1}) = B$

(ii) For every $b \in B$, if $f^{-1}(b) = a_1$ and $f^{-1}(b) = a_2$, then $a_1 = a_2$. That is, f^{-1} assigns exactly one value to each element in B.

From Corollary 2, if f is onto, then $Dom(f^{-1}) = B$; thus, f^{-1} has property (i), if f is onto. The next theorem gives the property of f so that (ii) is true for f^{-1}.

Theorem 3 For a function $f : A \to B$, if f is one-to-one, then

$$f^{-1}(b) = a_1 \text{ and } f^{-1}(b) = a_2 \text{ implies } a_1 = a_2 .$$

Proof

Assume that $f : A \to B$ is a one-to-one function. Then $f^{-1}(b) = a_1$ and $f^{-1}(b) = a_2$ means that (b,a_1) and (b,a_2) are ordered pairs in the relation f^{-1} from B to A. By definition of f^{-1}, it follows that (a_1,b) and (a_2,b) are ordered pairs in the function f from A to B. In function notation, this means that $f(a_1) = b$ and $f(a_2) = b$. Because, f is a one-to-one function, we have that $a_1 = a_2$. This shows that $f^{-1}(b) = a_1$ and $f^{-1}(b) = a_2$ implies $a_1 = a_2$, when f is a one-to-one function. ■

Definition 4 For a function $f : A \to B$, f **is invertible** means that f^{-1} is a function.

The following theorem summarizes these results.

Theorem 4 For a function $f : A \to B$, f is invertible if and only if f is one-to-one and onto.

Example 5 Which of the following functions from $\mathbb{R}$ to $\mathbb{R}$ are invertible?

(a) $f(x) = x^2$ (b) $f(x) = x^3$ (c) $f(x) = \frac{1}{x}$

Solution

(a) $f(x) = x^2$ is not one-to-one; thus, it is not invertible.
(b) $f(x) = x^3$ is one-to-one and onto; thus, it is invertible.
(c) Although $f(x) = \frac{1}{x}$ is one-to-one, it is not onto $\mathbb{R}$, since $\frac{1}{x} \neq 0$ for $x \in \mathbb{R}$. Therefore, $f(x)$ is not invertible.
Technically, the domain of $f(x) = \frac{1}{x}$ is $\mathbb{R} - \{0\}$ and $f(x) = \frac{1}{x}$ is one-to-one and onto from $\mathbb{R} - \{0\}$ to $\mathbb{R} - \{0\}$ Thus, it is invertible on the domain $\mathbb{R} - \{0\}$. ■

General Properties of Relations

If $x,y \in \mathbb{R}$, the relation $x = y$ has three important properties, which are listed as follows.

(!) $x = x$ for each $x \in \mathbb{R}$.
(2) If $x = y$, then $y = x$.
(3) If $x = y$ and $y = z$, then $x = z$.

Observe that for $x,y \in \mathbb{R}$, the relation $x < y$ does not satisfy (1) and (2); however, (3) is true for the $<$ relation.
We now give a general definition of these three properties.

In the following definitions, A is a nonempty set and $\sim$ is a relation on A.

Definition 5 The relation $\sim$ is **reflexive** provided that $x \sim x$ for each $x \in A$. If the relation is given as a set R of ordered pairs, then $(x,x) \in R$ for each $x \in A$.

Definition 6 The relation $\sim$ is **symmetric** on A provided that for $x,y \in A$, if $x \sim y$, then $y \sim x$.

Definition 7 The relation $\sim$ is **transitive** on A provided that for $x,y,z \in A$, if $x \sim y$ and $y \sim z$, then $x \sim z$.

Example 6

Let A be the set of all lines in the plane. For l_1, l_2 any two lines in A, define $l_1 \sim l_2$ by $l_1 \perp l_2$.

Observe that $l_1 \perp l_2$ and $l_2 \perp l_3$ will make l_1 and l_3 parallel lines; thus, $\perp$ is not transitive.
Since a line can not be perpendicular to itself, $\sim$ is not reflexive. If $l_1 \perp l_2$, then $l_2 \perp l_1$ so that $\perp$ is a symmetric relation. ■

Example 7

Let U be a finite, nonempty set and let $P(U)$ be the set of all subsets of U. For $A,B \in P(U)$, $A \sim B$ if and only if $A \cap B \neq \varnothing$.

Because $\varnothing \cap \varnothing = \varnothing$ we have $\varnothing \nsim \varnothing$; thus, $\sim$ is not reflexive.

Let $A,B \in P(U)$. If $A \cap B \neq \varnothing$, then $B \cap A \neq \varnothing$, since $A \cap B = B \cap A$. Thus, $\sim$ is symmetric.

To show that $\sim$ is not transitive, we must construct a counterexample, that is, we must find three sets A,B,C such that $A \cap B \neq \varnothing$ and $B \cap C \neq \varnothing$, yet $A \cap C = \varnothing$.
Consider $U = \{1,2,3,4,5,6,7,8,9,10\}$, $A = \{1,2,3,4\}$, $B = \{3,4,5,6\}$, $C = \{5,6,7,8\}$, then $A \cap B \neq \varnothing$ and $B \cap C \neq \varnothing$, yet $A \cap C = \varnothing$. ■

Example 8

For $x,y \in \mathbb{R} - \{0\}$, let $x \sim y$ if and only if $\frac{x}{y} \in \mathbb{Q}$.

Note that if $x = \pi$ and $y = 1$ then $\frac{x}{y} = \frac{\pi}{1} = \pi \notin \mathbb{Q}$. This illustrates that there are many cases where $\frac{x}{y} \notin \mathbb{Q}$ for $x,y \in \mathbb{R} - \{0\}$.
For all $x \in \mathbb{R} - \{0\}$, we have $\frac{x}{x} = 1 \in \mathbb{Q}$ and therefore $x \sim x$. Thus, $\sim$ is reflexive.
If $x \sim y$ then $\frac{x}{y} \in \mathbb{Q}$. Because $\mathbb{Q}$ is a field and $\frac{x}{y} \neq 0$, then $(\frac{x}{y})^{-1} = \frac{y}{x} \in \mathbb{Q}$. Thus, $y \sim x$ follows from $x \sim y$ and $\sim$ is symmetric.
If $x \sim y$ and $y \sim z$, then $\frac{x}{y} \in \mathbb{Q}$ and $\frac{y}{z} \in \mathbb{Q}$. Therefore $(\frac{x}{y})(\frac{y}{z}) = \frac{x}{z} \in \mathbb{Q}$ and $x \sim z$, since $\mathbb{Q}$ is closed for multiplication. This shows that $x \sim z$ follows from $x \sim y$ and $y \sim z$. Thus, $\sim$ is transitive. Therefore, $\sim$ is reflexive, symmetric, and transitive. ■

Example 9

For $x,y \in \mathbb{Z} - \{0\}$, let $x \sim y$ if and only if $x \mid y$.

Since $x \cdot 1 = x$ is true for each $x \in \mathbb{Z} - \{0\}$, we have $x \mid x$ and thereby $x \sim x$. Thus, $\sim$ is reflexive.
If $x \mid y$ and $y \mid z$, then $x \mid z$ follows. Thus, $x \sim y$ and $y \sim x$ imply that $x \sim z$. Thus, $\sim$ is transitive.
Let $x = 2$ and $y = 6$, then it is true that $x \mid y$ and false that $y \mid x$. That is, $2 \sim 6$ yet $6 \nsim 2$.
Consequently, $\sim$ is not symmetric. ■

Equivalence Relations

The previous examples illustrate that many relations do not satisfy all three properties - reflexive,symmetric, and transitive. If a relation does have these three properties then it can be shown that other desirable results follow. We now make the following definition.

Definition 8 A relation $\sim$ on a nonempty set A is an **equivalence** relation provided that $\sim$ is reflexive, symmetric, and transitive. If $\sim$ is an equivalence relation and $x \sim y$, then we say that x is equivalent to y.

Example 10

For $\frac{x}{y}, \frac{a}{b} \in \mathbb{Q}$ define $\frac{x}{y} \sim \frac{a}{b}$ if and only if $xb = ya$.

It can be shown that $\sim$ is an equivalence relation (see Exercise 10).
Observe that $\frac{1}{2} \sim \frac{2}{4} \sim \frac{3}{6}$ and so forth. Also, $\frac{2}{3} \sim \frac{4}{6} \sim \frac{8}{12}$ and so forth. These rational numbers illustrate that $\frac{x}{y} \sim \frac{a}{b}$ iff $\frac{x}{y} = \frac{a}{b}$.
Moreover, we can group rational numbers into subsets so that all the rational numbers in a given subset are equivalent to each other. These subsets are called equivalence classes. ■

We now study equivalence classes in more detail.

Definition 9 Let $\sim$ be an equivalence relation on a nonempty set A. For $x \in A$, the **equivalence class** of x determined by $\sim$ is the subset of A, denoted by $[x]$, consisting of all elements of A that are equivalent to x, that is

$$[x] = \{a \in A \mid a \sim x\}$$

$[x]$ is called the "equivalence class" of x.

Using $\sim$ as given in Example 10, $[\frac{1}{2}] = \{\frac{1}{2}, \frac{2}{4}, \frac{3}{6}, \frac{4}{8}, \ldots\}$ and $[\frac{2}{3}] = \{\frac{2}{3}, \frac{4}{6}, \frac{8}{12}, \frac{12}{18}, \ldots\}$.

For an equivalence relation on a set A, every element of A is in its own equivalence class. Moreover, two equivalence classes are either identical or they have no elements in common. These properties are proved in the next theorem.

Theorem 5 Let $\sim$ be an equivalence relation on a nonempty set A, then

(1) For each $a \in A$, $a \in [a]$.
(2) For $a, b \in A$, $a \sim b$ if and only if $[a] = [b]$.
(3) For $a, b \in A$, either $[a] = [b]$ or $[a] \cap [b] = \varnothing$.

Proof:
To prove (1) we observe that for each $a \in A$, $a \sim a$ since $\sim$ is an equivalence relation and has the reflexive property. Thus, $a \in [a]$ by definition of $[a]$.

To prove (2), we must prove each of the following statements:
(2a) If $a \sim b$, then $[a] = [b]$.
(2b) If $[a] = [b]$, then $a \sim b$.

To prove (2a), we assume that $a \sim b$, then prove that $[a]=[b]$ by proving that $[a] \subseteq [b]$ and $[b] \subseteq [a]$. Suppose that $x \in [a]$, then $x \sim a$. Thus, $x \sim a$ and $a \sim b$ are both true so that $x \sim b$ by the transitive property of $\sim$. Therefore, $x \in [b]$. Thus, $[a] \subseteq [b]$ follows.
Next, we show that $[b] \subseteq [a]$. Suppose that $y \in [b]$, then $y \sim b$. Moreover, $b \sim a$ by symmetry since $a \sim b$. Thus, we have that $y \sim b$ and $b \sim a$ are both true so that $y \sim a$ by the transitive property of $\sim$. Therefore, $y \in [a]$. Thus, $[b] \subseteq [a]$ follows. This completes the proof of (2a).

To prove (2b), we assume that $[a]=[b]$. Because $a \in [a]$ from (1) and $[a]=[b]$, we have that $a \in [b]$. Thus, $a \sim b$ by definition of $[b]$. This completes the proof of (2b).

To prove (3), we use the fact that either $[a] \cap [b] = \varnothing$ or $[a] \cap [b] \neq \varnothing$.
We can show that when $[a] \cap [b] \neq \varnothing$, then $[a]=[b]$.
Suppose that $[a] \cap [b] \neq \varnothing$ and let $x \in [a] \cap [b]$, then $x \in [a]$ and $x \in [b]$. Therefore, $x \sim a$ and $x \sim b$. By symmetry, this means that $a \sim x$ and $x \sim b$. From the transitive property, this implies that $a \sim b$. Consequently, $[a]=[b]$ by result (2) of this theorem. Thus, whenever $[a] \cap [b] \neq \varnothing$ then $[a]=[b]$. Therefore, we have shown that either $[a] \cap [b] = \varnothing$ or $[a]=[b]$. ■

The following table states each result of Theorem 5 and gives a corresponding verbal description.

Mathematical statement	Verbal Description
For each $a \in A$, $a \in [a]$	Every element of A is in its own equivalence class.
For $a,b \in A$, $a \sim b$ iff $[a]=[b]$	Two elements of A are equivalent if and only if their equivalence classes are equal.
For $a,b \in A$, $[a]=[b]$ or $[a] \cap [b] = \varnothing$	Any two equivalence classes are either equal or disjoint. If two equivalence classes have at least one element in common, then they are equal.

Table 1

Partitions

Result (3) of Theorem 5 shows that the collection of all equivalence classes determined by an equivalence relation on a nonempty set A is a collection of subsets of A, wherein each pair of distinct subsets is disjoint. Moreover, for each $a \in A$, $a \in [a]$. Consequently, the collection of all equivalence classes creates what is called a partition of A. The following is formal definition.

Definition 10 Let A be a nonempty set and let $\mathbb{C}$ be a collection of subsets of A. $\mathbb{C}$ is a **partition** of A provided that
(a) For each $U \in \mathbb{C}$, $U \neq \varnothing$.
(b) For each $x \in A$, there exists a $U \in \mathbb{C}$ such that $x \in U$.
(c) For every $U,V \in \mathbb{C}$, we have $U=V$ or $U \cap V = \varnothing$.

(c) states that if U,V are any two subsets in a partition, then they are either equal or disjoint.

Examples of Partitions

(i) $A = \{1,2,3,4,5\}$ and $\mathbb{C} = \{\{1\},\{2,3\},\{4,5\}\}$.
(ii) $A = \mathbb{Z}$ and $U = \{$all even integers$\}$ and $V = \{$all odd integers$\}$, then $\mathbb{C} = \{U, V\}$.
(iii) $A = \{a,b,c,\ldots x,y,z\}$ and $U = \{$all vowels$\}$ and $V = \{$all consonants$\}$, then $\mathbb{C} = \{U, V\}$.

The next theorem uses Theorem 5 to prove that given an equivalence relation on a set A , the collection of all equivalence classes form a partition of A .

Theorem 6 Let $\sim$ be an equivalence relation on a nonempty set A . The collection $\mathbb{C}$ of all equivalences classes forms a partition of A .

Proof We need to show that the collection of equivalence classes satisfies conditions (a), (b), (c) of the definition of a partition.

(a) and (b) follow from the facts that for each $a \in A$, $a \in [a]$ and $[a] \neq \varnothing$.

Next, let $U, V \in \mathbb{C}$, then there exist $a, b \in A$ such that $U = [a]$ and $V = [b]$. From (3) of Theorem 5, either $[a] = [b]$ and hence $U = V$ or $[a] \cap [b] = \varnothing$, in which case $U \cap V = \varnothing$. ■

Making Relations From Partitions

Can we reverse the process of creating a partition by using the equivalence classes for a certain equivalence relation? In other words, if we have a partition $\mathbb{C}$ on a set A , is there an equivalence relation on A such that its equivalence classes are precisely the sets in the partition $\mathbb{C}$?

To answer this question affirmatively, we begin by describing how a partition can be used to define a equivalence relation.

Definition 11 For a set A and a partition $\mathbb{C}$ of A , the relation associated with $\mathbb{C}$ (denoted $\sim_{\mathbb{C}}$) is the relation defined by $x \sim_{\mathbb{C}} y$ if and only if there exists a $U \in \mathbb{C}$ such that $x \in U$ and $y \in U$.

Example 11 Let $A = \{1,2,3,4,5\}$ and $\mathbb{C} = \{\{1,2\},\{3,4\},\{5\}\}$. We can list the ordered pairs (x,y) such that $x \sim_{\mathbb{C}} y$ as follows:

$$R = \{(1,1),(1,2),(2,1),(2,2),(3,3),(3,4),(4,3),(4,4),(5,5)\}$$

Notice that R includes the pairs $(1,1),(2,2),(3,3),(4,4)$, and $(5,5)$ because we can use $x = y$ in $x \sim_{\mathbb{C}} y$. Thus, R is reflexive.
Observe also that if $(x,y) \in R$, then $(y,x) \in R$. Thus, $x \sim_{\mathbb{C}} y$ is a symmetric relation. We can also show that $x \sim_{\mathbb{C}} y$ is transitive. ■

We generalize the results of this example by the following theorem.

Theorem 7 Let A be a nonempty set and $\mathbb{C}$ be a collection of subsets of A. If $\mathbb{C}$ is a partition of A, then $x \sim_{\mathbb{C}} y$ is reflexive, symmetric, and transitive.

Proof

To show $\sim_{\mathbb{C}}$ is reflexive, let $x \in A$. Because $\mathbb{C}$ is a partition, there exists $U \in \mathbb{C}$ such that $x \in U$. Thus, $x \sim_{\mathbb{C}} x$ by definition.

Suppose that $x \sim_{\mathbb{C}} y$, then there exist $U \in \mathbb{C}$ with $x \in U$ and $y \in U$. This means that $y \in U$ and $x \in U$ so that $y \sim_{\mathbb{C}} x$. Thus, $\sim_{\mathbb{C}}$ is symmetric.

To show that $\sim_{\mathbb{C}}$ is transitive, we must show that if $x \sim_{\mathbb{C}} y$ and $y \sim_{\mathbb{C}} z$, then $x \sim_{\mathbb{C}} z$ follows. Assume that $x \sim_{\mathbb{C}} y$ and $y \sim_{\mathbb{C}} z$, then there are sets U and V in the partition such that $x,y \in U$ and $y,z \in V$. Since $y \in U$ and $y \in V$, we have that $U \cap V \neq \varnothing$. From condition (c) in the definition of a partition, we must have $U = V$. Therefore, $x,y,z \in U = V$. Consequently, $x \sim_{\mathbb{C}} z$ from the definition of $\sim$. ■

Theorem 7 shows that $\sim_{\mathbb{C}}$ is an equivalence relation.
The next theorem shows that if $\mathbb{C}$ is a partition of A, then the equivalence classes of $x \sim_{\mathbb{C}} y$ are precisely the sets in $\mathbb{C}$.

Theorem 8 For each $a \in A$, let $U \in \mathbb{C}$ be the set such that $a \in U$, then $[a] = U$ for the equivalence relation $\sim_{\mathbb{C}}$.

Proof Let U be a set in $\mathbb{C}$ and assume that $a \in U$, we must show that $[a] = U$.
Let $x \in [a]$, then $x \sim_{\mathbb{C}} a$ and therefore $\exists\ V \in \mathbb{C}$ such that $x,a \in V$. Since $a \in U$ and $a \in V$, we have that $U \cap V \neq \varnothing$. Consequently, $U = V$ since U, V are sets in a partition. Thus, $x \in U = V$ and $[a] \subseteq U$.

Now let $y \in U$. Since $a,y \in U$, we have $y \sim_{\mathbb{C}} a$ and thus $y \in [a]$ by definition of an equivalence class. This shows that $U \subseteq [a]$.
Because $[a] \subseteq U$ and $U \subseteq [a]$, it follows that $U = [a]$. ■

Theorem 8 shows that every set U in a partition $\mathbb{C}$ for A is equal to the equivalence class $[a]$ for some $a \in A$. Also every equivalence class $[a]$ equals the set U in $\mathbb{C}$ for which $a \in U$. Thus, the sets in $\mathbb{C}$ are the equivalence classes created by the equivalence relation $\sim_{\mathbb{C}}$.

EXERCISES

1. Let $A = \{x,y,z\}, B = \{a,b,c\}$, and $R = \{(x,a),(y,a),(y,b),(y,c)\}$

(a) Use the roster method to list all the elements of $A \times B$.
(b) Explain why R is a relation from A to B.
(c) Find the domain and image of R.
(d) Determine R^{-1}, the inverse relation of R.

2. Let $A = \{x,y,z\}$ and $R = \{(x,x),(y,y),(x,y),(y,x)\}$

(a) Is R a reflexive relation? Explain.
(b) Is R a symmetric relation? Explain.
(c) Is R a transitive relation? Explain.

3. Let $R = \{(x,y) \in \mathbb{R} \times \mathbb{R} \mid x^2 + y^2 = 25\}$. Note that R is a relation on $\mathbb{R}$.

(a) Find the domain of R .
(b) Find the image of R .
(c) Is R a function from $\mathbb{R}$ to $\mathbb{R}$?
(d) Graph the elements of R in the coordinate plane. Is the graph consistent with your answers for (a)-(c) ?

4. Prove the following:
" Let A and B be nonempty sets and R and S be relations from A to B . If $R \subseteq S$, then $R^{-1} \subseteq S^{-1}$."

5. If R is a relation from A to B , prove

(a) $\text{Im}(R^{-1}) = Dom(R)$
(b) $(R^{-1})^{-1} = R$

6. Determine which of the following functions from $\mathbb{R}$ to $\mathbb{R}$ is invertible.

(a) $f(x) = 3x + 5$ (b) $f(x) = x^4$ (c) $f(x) = x^5$

7. Let $A = \{w,x,y,z\}$. Give an example of a relation on A that has none of the properties: reflexive; symmetric; transitive.

8. Define a relation on $\mathbb{Z}$ by $x \sim y$ iff $x \cdot y \geq 0$. Prove or disprove the following:

(a) $\sim$ is reflexive (b) $\sim$ is symmetric (c) $\sim$ is transitive

For transitive, use the fact that if $x \cdot y \geq 0$, then ($x \geq 0$ and $y \geq 0$) or ($x < 0$ and $y < 0$).

9. Define a relation on $\mathbb{Z}$ by $x \sim y$ iff $x + y$ is even.

(a) Show that $\sim$ is an equivalence relation.
(b) Find the distinct equivalence classes.

Hint: For transitive, consider various cases where x,y,z are even or odd.

10. Consider the properties reflexive,symmetric, and transitive. For each of the following relations, state which of the of the preceding properties it has and justify your answer. If it does not have a

certain property, give a counterexample.

(a) Let U be a nonempty, finite set, with subsets $A, B \in P(U)$. Define $A \sim B$ iff $A \cap B = \varnothing$

(b) $\sim$ defined on $\mathbb{R}$ by $x \sim y$ iff $x - y \in \mathbb{Z}$

(c) Let $f : \mathbb{R} \to \mathbb{R}$ be a function defined $f(x) = x^2 + 5$. Define $\sim$ on $\mathbb{R}$ by $x \sim y$ iff $f(x) = f(y)$.

(d) $\sim$ defined on $\mathbb{Z}$ by $x \sim y$ iff $x + y$ is odd.

(e) $\sim$ defined on the set of all lines in the plane by $l_1 \sim l_2$ iff l_1 and l_2 intersect in a point.

(f) $\sim$ defined on $\mathbb{Q}$ by $\frac{x}{y} \sim \frac{a}{b}$ iff $xb = ay$.

11. Let $\sim$ be an equivalence relation on $A = \{a, b, c, d\}$. Prove that if $a \sim b$, $c \sim d$ and $a \sim d$, then $b \sim c$.

12. Let $S \subseteq \mathbb{Z}$ and for $x, y \in S$ define $x \sim y$ iff $3 \mid (x + 2y)$.

(a) Prove that $\sim$ is an equivalence relation.
(b) For $S = \{0, 1, 2, 3, 4, 5, 6, 7\}$, find the distinct equivalence classes of S for $\sim$.

13. Define $\sim$ on $\mathbb{N}$ by $m \sim n$ iff

$$\{p \mid p \text{ is prime and } p \mid m\} = \{p \mid p \text{ is prime and } p \mid n\}$$

Thus, $m \sim n$ iff m and n have the same prime divisors.

(a) Find two specific natural numbers x and y such that $x, y \in [15]$.
(b) Find two specific natural numbers x and y such that $x, y \in [24]$.

14. Define $\sim$ on $\mathbb{R} \times \mathbb{R}$ by $(a, b) \sim (c, d)$ iff $a^2 + b^2 = c^2 + d^2$.

(a) Show that $\sim$ is an equivalence relation.
(b) Use the fact that the distance d from $(0, 0)$ to (a, b) satisfies $d^2 = a^2 + b^2$ to describe the set of points in $\mathbb{R} \times \mathbb{R}$ that are in the equivalence class of (a, b) .

15. Determine whether each of the following collection of sets $\mathbb{C}$ is a partition of the indicated set. If it is not a partition, then justify why a particular condition in Definition 10 fails to be satisfied.

(a) $A = \mathbb{R}$, $\mathbb{C} = \{[n, n+1) \mid n \in \mathbb{Z}\}$, where each $[n, n+1)$ is the interval $\{x \mid n \le x < n+1\}$.

(b) $A = \mathbb{R} \times \mathbb{R}$, $\mathbb{C}$ is the set of all lines with slope 2 .

(c) $A = \mathbb{Z}$, $\mathbb{C} = \{E, O\}$ where E is the set of all even integers and O is the set of all odd integers.

(d) $A = \mathbb{R} \times \mathbb{R}$, $\mathbb{C} = \{[n, n+1) \times \mathbb{R} \mid n \in \mathbb{Z}\}$

16. Suppose that A and B are sets with partitions $\mathbb{C}_A$ and $\mathbb{C}_B$ respectively. Prove that " if $A \cap B = \varnothing$, then $\mathbb{C}_A \cup \mathbb{C}_B$ is a partition on $A \cup B$ " .

17. List the elements in each set of the partition of $A = \{a, b, c, d, e\}$ that corresponds to the given equivalence relation for each of the following.

(a) $R_1 = \{(a,a),(b,b),(c,c),(d,d),(e,e),(a,b),(b,a),(c,d),(d,c)\}$
(b) $R_2 = \{(a,a),(b,b),(c,c),(d,d),(e,e)\}$
(c) $R_3 = \{(a,a),(b,b),(c,c),(d,d),(e,e),(a,e),(e,a)\}$
Hint: First, find the equivalence classes for each relation.

18. Describe, by listing all ordered pairs, the equivalence relation on the set $\{a, b, c, d, e\}$ corresponding to each of the following partitions:

(a) $\mathbb{C}_1 = \{\{a,b,c\},\{d,e,f\}\}$
(b) $\mathbb{C}_2 = \{\{a\},\{b\},\{c,d\{,\{e,f\}\}$
(c) $\mathbb{C}_3 = \{\{a,b\},\{c,d\},\{e,f\}\}$

19. Let $f : \mathbb{R} \to \mathbb{R}$ be a function. Show that $\sim$ defined on $\mathbb{R}$ by $a \sim b$ iff $f(a) = f(b)$ is an equivalence relation.

20. Define $\sim$ on $\mathbb{Z}$ by $m \sim n$ iff $2 \mid (3m - n)$. Prove $\sim$ is an equivalence relation on $\mathbb{Z}$. Determine the distinct equivalence classes for $\sim$.

CHAPTER 8 LIMITS AND CONTINUITY

Limits - Definitions and Theorems

Let $f: \mathbb{R} \to \mathbb{R}$ be a function ; here the actual domain of f may only be a proper subset of $\mathbb{R}$. Intuitively, we say that a real number L is the limit value of $f(x)$ as x approaches the value a, if the values of $f(x)$ can be made arbitrarily close to L by making x sufficiently close to a. We write $\lim_{x \to a} f(x) = L$ or $\lim_{x \to a} f(x) = L$. Here, $f(a)$ may not be defined; however, we need $f(x)$ defined for an interval around a, which can be described as $x \in (a-\delta, a) \cup (a, a+\delta)$ for some $\delta > 0$.

Example 1 Let $f(x) = \frac{x^2-4}{x-2}$. Set up a table of values for x close to $a = 2$ to determine the $\lim_{x \to 2} f(x) = L$.

Solution We use the following table. Observe that $f(x)$ is not defined (ND) for $x = 2$.

x	1.9	1.99	1.999	2	2.0001	2.01	2.1
$f(x)$	3.9	3.99	3.999	ND	4.0001	4.01	4.1

L appears to be 4, even though $f(x)$ is not defined for $x = 2$. We will verify this later on. ■

The desired closeness of $f(x)$ to L is generally denoted by "ϵ" (the Greek letter epsilon). The Greek letter "δ" (delta) is usually used to denote how close x must be to a to achieve the desired closeness of $f(x)$ to L. The following diagram illustrates this.

```
       a − δ       a         x        a + δ
__________(________)(_______•_________)______
                            ↓ f
__________(_________________•_________)_________
     L − ϵ        L       f(x)       L + ϵ
```

In other words, to show that $\lim_{x \to a} f(x) = L$ requires that given any $\epsilon > 0$, we must be able to find a $\delta > 0$ such that the set $(a-\delta, a) \cup (a, a+\delta)$ is mapped by f into the interval $(L-\epsilon, L+\epsilon)$. Since the statement $x \in (a-\delta, a) \cup (a, a+\delta)$ is equivalent to the statement $0 < | x - a | < \delta$ and the statement $f(x) \in (L-\epsilon, L+\epsilon)$ is equivalent to $| f(x) - L | < \epsilon$, we make the following definition.

Definition 1 For a function $f: \mathbb{R} \to \mathbb{R}$, the " limit of $f(x)$ as x approaches a is L", provided that for each $\epsilon > 0$, there exists a $\delta > 0$ such that $0 < | x - a | < \delta$ implies $| f(x) - L | < \epsilon$. We write that $\lim_{x \to a} f(x) = L$.

In Definition 1, the condition $0 < | x - a | < \delta$ allows for the possibility that $f(x)$ may not be defined for $x = a$. Example 1 gives a function that illustrates this possibility.

Example 2 Let $f(x) = 2x + 5$ and use Definition 1 to verify that $\lim_{x \to 4} f(x) = 13$.

Solution According to Definition 1, given a positive value for ϵ, we need to find a positive value for δ such that if $0 < | x - 4 | < \delta$, then $| 2x + 5 - 13 | < \epsilon$, that is, $| 2x - 8 | < \epsilon$.
We observe that

$$| 2x - 8 | = | 2(x - 4) | = | 2 | | x - 4 | = 2 | x - 4 |$$

Thus, if $0 < | x - 4 | < \frac{\epsilon}{2}$, then $| 2x - 8 | = 2 | x - 4 | < 2(\frac{\epsilon}{2}) = \epsilon$. Thus, $\delta = \frac{\epsilon}{2}$ works. ■

Example 3 Use Definition 1 to verify that $\lim_{x \to 3} x^2 = 9$.

Solution According to definition 1, given a positive value for ϵ, we need to find a positive value for δ such that if $0 < | x - 3 | < \delta$, then $| x^2 - 9 | < \epsilon$. We observe that

$$| x^2 - 9 | = | x + 3 | | x - 3 |$$

Also, if x is close to 3, then $| x + 3 |$ is close to 6. For example, if we let $\delta = .5$, then $0 < | x - 3 | < .5$ gives $3 - .5 < x < 3 + .5$, that is, $2.5 < x < 3.5$ and hence $5.5 < x + 3 < 6.5$ so that $| x + 3 | < 6.5$.
Suppose that $\epsilon > 0$ and let $\delta =$ minimum of $\{ \frac{\epsilon}{6.5}, .5\}$. Then, $0 < | x - 3 | < \delta$ means that $0 < | x - 3 | < \frac{\epsilon}{6.5}$ and $| x + 3 | < 6.5$ (because $\delta \leq .5$). Therefore,

$$| x^2 - 9 | = | x + 3 | | x - 3 | < 6.5(\frac{\epsilon}{6.5}) = \epsilon$$ ■

We can prove (Exercise 11) that $\lim_{x \to a} f(x)$ gives a unique value and derive the next theorem.

Theorem 1 If $\lim_{x \to a} f(x) = L_1$ and $\lim_{x \to a} f(x) = L_2$, then $L_1 = L_2$.

Most functions are too complicated to use Definition 1 to compute limits. However, most functions can be expressed as a sum, difference, multiplication, or division of simpler functions. For example $f(x) = x^3 + 3x$ can be written as the sum of $g(x) = x^3$ and $h(x) = 3x$. Thus, computing limits can be done by computing limits for simpler functions. We now present some theorems to implement this approach. *We are assuming that* $f: \mathbb{R} \to \mathbb{R}$ *is a function whose domain is a subset of* $\mathbb{R}$.

Theorem 2 If $\lim_{x \to a} f(x) = L$ and $\lim_{x \to a} g(x) = M$, then

(a) $\lim_{x \to a} [f(x) + g(x)] = L + M$

(b) $\lim_{x \to a} [f(x) - g(x)] = L - M$

Proof To prove (a), we must show that for each $\epsilon > 0$, $| f(x) + g(x) - (L + M) | < \epsilon$, if $0 < | x - a | < \delta$ for a suitable choice of $\delta > 0$. The key idea is that

$$| f(x) + g(x) - (L + M) | = | (f(x) - L) + (g(x) - M) | \leq | f(x) - L | + | g(x) - M |$$

If we can choose $\delta > 0$ so that both $| f(x) - L | < \frac{\epsilon}{2}$ and $| g(x) - M | < \frac{\epsilon}{2}$, then from (2-i)

$$| f(x) + g(x) - (L + M) | < \frac{\epsilon}{2} + \frac{\epsilon}{2} = \epsilon$$

First, since $\lim_{x \to a} f(x) = L$, choose $\delta_1 > 0$ so that if $0 < | x - a | < \delta_1$, then $| f(x) - L | < \frac{\epsilon}{2}$.
Next, since $\lim_{x \to a} g(x = M$, choose $\delta_2 > 0$ so that if $0 < | x - a | < \delta_2$, then $| g(x) - M | < \frac{\epsilon}{2}$.

Now, choose $\delta = \min\{\delta_1, \delta_2\}$ and let $0 < | x - a | < \delta$. It follows that

$$0 < | x - a | < \delta_1 \text{ and } 0 < | x - a | < \delta_2$$

Therefore,

$$| f(x) - L | < \tfrac{\epsilon}{2} \text{ and } | g(x) - M | < \tfrac{\epsilon}{2}$$

Consequently, from (2-i)

$$| f(x) + g(x) - (L + M) | < \frac{\epsilon}{2} + \frac{\epsilon}{2} = \epsilon$$ ■

The proof of (b) is left as an exercise.

Theorem 2 states that the limit of the sum (or difference) of two functions is the sum (or difference) of their limits.

Theorem 3 If $\lim_{x\to a} f(x) = L$ and c is a nonzero constant, then $\lim_{x\to a}[cf(x)] = c(L)$.

Proof We must show that for any $\epsilon > 0$, that $| c[f(x)] - c(L) | < \epsilon$, if $0 < | x - a | < \delta$ for a suitable choice of $\delta > 0$. The following equality gives a method for choosing δ.

$$| c[f(x)] - c(L) | = | c[f(x) - (L)] | = | c | | f(x) - L | \qquad (3i)$$

Since $\lim_{x\to a} f(x) = L$ and $\frac{\epsilon}{|c|} > 0$, we can choose $\delta > 0$ so that $| f(x) - L | < \frac{\epsilon}{|c|}$, whenever $0 < | x - a | < \delta$.

Therefore, by (3i), the following inequality

$$| c[f(x)] - c(L) | = | c | | f(x) - L | < | c | \frac{\epsilon}{| c |} = \epsilon$$

will be true for $0 < | x - a | < \delta$. ■

Theorem 4 If $\lim_{x\to a} f(x) = L$ and $\lim_{x\to a} g(x) = M$, then $\lim_{x\to a} f(x) \cdot g(x) = L \cdot M$.

Sketch of proof The key idea is that

$$| f(x) \cdot g(x) - L \cdot M | = | f(x) \cdot g(x) - f(x) \cdot M + f(x) \cdot M - L \cdot M | = | f(x) \cdot [g(x) - M] + [f(x) - L] \cdot M |$$

$$\leq | f(x) | | g(x) - M | + | f(x) - L | | M |$$

If we can make each of $| f(x) | | g(x) - M |$ and $| f(x) - L | | M |$ less than $\frac{\epsilon}{2}$ then

$$| f(x) \cdot g(x) - L \cdot M | < \frac{\epsilon}{2} + \frac{\epsilon}{2} = \epsilon$$

Observe that $| M |$ is a nonnegative constant and both $| g(x) - M |$ and $| f(x) - L |$ can be made arbitrarily small if $0 < | x - a | < \delta$ for a suitable choice of $\delta > 0$, since $\lim_{x\to a} f(x) = L$ and $\lim_{x\to a} g(x) = M$.

We need to show that $| f(x) | \leq B$ for some $B > 0$, when $0 < | x - a | < \delta$. Let $\epsilon = 1$, then $\exists$ $\delta > 0$ such that if $0 < | x - a | < \delta$, then $| f(x) - L | < 1$ because $\lim_{x\to a} f(x) = L$. Thus,

$$| f(x) | = | f(x) - L + L | \leq | f(x) - L | + | L | < 1 + | L |$$

Therefore, we can use $B = 1 + | L |$. The remaining details of the proof are left as an exercise. ■

We also have the following useful theorems.

Theorem 5 If $\lim_{x \to a} f(x) = L$ and $\lim_{x \to a} g(x) = M$, where $M \neq 0$, then $\lim_{x \to a} \frac{f(x)}{g(x)} = \frac{L}{M}$.

Theorem 6 Let $a, c \in \mathbb{R}$. If $f(x) = c$ for each $x \in \mathbb{R}$, then $\lim_{x \to a} f(x) = c$.

The proofs of Theorem 5 and Theorem 6 are left as exercises.

Theorem 7 Let $f(x) = x$ for each $x \in \mathbb{R}$. Then for each $a \in \mathbb{R}$, $\lim_{x \to a} f(x) = \lim_{x \to a} x = a$.

Proof Let $\epsilon > 0$ be given. Since $| f(x) - a | = | x - a |$, we can make $| f(x) - a | < \epsilon$ simply by choosing $\delta = \epsilon$. In this case, if $0 < | x - a | < \delta = \epsilon$, then $| f(x) - a | = | x - a | < \epsilon$. ■

The next two theorems generalize some of the previous theorems.

Theorem 8 Let $n \in \mathbb{N}$ and $f(x) = x^n$ for all $x \in \mathbb{R}$, then $\lim_{x \to a} f(x) = \lim_{x \to a} x^n = a^n$.

Proof We use the method of Mathematical Induction.
For $n = 1$, $f(x) = x$; thus, $\lim_{x \to a} f(x) = a = a^1$, which is true by Theorem 7. This verifies P(1) .
Assume for $k \geq 1$ that P(k) is true, that is, $\lim_{x \to a} x^k = a^k$. We must show that P$(k+1)$ is true, that is, $\lim_{x \to a} x^{k+1} = a^{k+1}$.
By the rules of exponents, $x^{k+1} = x(x^k)$. Then by Theorem 4 and Theorem 7, we have

$$\lim_{x \to a} x^{k+1} = \lim_{x \to a} x(x^k) = [\lim_{x \to a} x][\lim_{x \to a} x^k] = a[a^k] = a^{k+1}$$ ■

We can also use Mathematical Induction to prove the next theorem (see Exercise 4).

Theorem 9 Let $f_1, f_2, \ldots, f_n$ be n functions such that

$$\lim_{x \to a} f_1(x) = L_1 \ , \ \lim_{x \to a} f_2(x) = L_2 \ , \ldots, \ \lim_{x \to a} f_n(x) = L_n$$

then $\lim_{x \to a} [f_1(x) + f_2(x) + \cdots + f_n(x)] = L_1 + L_2 + \cdots + L_n$.

Example 4 Show the use of appropriate theorems to compute $\lim_{x \to 5} [x^4 + 3x^3 - 2x]$.

Solution Using Theorem 9, we have

$$\lim_{x \to 5} [x^4 + 3x^3 - 2x] = \lim_{x \to 5} x^4 + \lim_{x \to 5} 3x^3 + \lim_{x \to 5} [-2x]$$

By Theorem 3, we have

$$\lim_{x \to 5} x^4 + \lim_{x \to 5} 3x^3 + \lim_{x \to 5} [-2x] = \lim_{x \to 5} x^4 + 3 \lim_{x \to 5} x^3 - 2 \lim_{x \to 5} x$$

Finally, from Theorem 8, we have

$$\lim_{x \to 5} x^4 + 3 \lim_{x \to 5} x^3 - 2 \lim_{x \to 5} x = 5^4 + 3[5^3] - 2[5] = 625 + 3[125] - 2[5] = 990$$

These theorems can be used to prove for a polynomial function $p(x) = c_n x^n + c_{n-1}x^{n-1} + \cdots + c_1 x + c_0$ that

$$\lim_{x \to a} p(x) = c_n a^n + c_{n-1}a^{n-1} + \cdots + c_1 a + c_0 = p(a)$$

Continuity

Intuitively, a function is continuous at $x = a$ provided that $f(a)$ is defined and there is no "break" between the pieces of the graph near $x = a$. We formalize this concept by the following definition.

Definition 3 For a function $f : \mathbb{R} \to \mathbb{R}$, with D the domain of f. Let $a \in D$ and $S \subseteq D$, then

(1) " f **is continuous at** a " means that $\lim_{x \to a} f(x) = f(a)$.
(2) " f **is continuous on** S " means that f is continuous at a, for each $a \in S$.

Observe that for a function f to be continuous at $x = a$, three properties are required:

1. $f(a)$ must be defined;
2. $\lim_{x \to a} f(x) = L$ must exist;
3. $L = f(a)$.

When f is continuous at $x = a$, then $f(a)$ is defined and $\lim_{x \to a} f(x) = f(a)$. Therefore, given any $\epsilon > 0$, there exists $\delta > 0$ such that $| x - a | < \delta$ implies $| f(x) - f(a) | < \epsilon$.
In this situation " $0 < | x - a | < \delta$ " can be replaced by " $| x - a | < \delta$ ". Using interval notation, we have for $\epsilon > 0$, that $x \in (a - \delta, a + \delta)$ implies $f(x) \in (f(a) - \epsilon , f(a) + \epsilon)$ for suitable choice of δ.

A wide variety of functions are continuous at all points in their domains. The previous results have shown that all polynomial functions are continuous. Moreover, all rational functions of the form $r(x) = \frac{f(x)}{g(x)}$ with $f(x)$ and $g(x)$ polynomial functions are continuous, except where $g(x) = 0$.
In order to work with functions which are continuous on sets of numbers, we need the concept of open sets in $\mathbb{R}$.

Definition 4 Let $S \subseteq \mathbb{R}$, then S is a nonempty **open set in** $\mathbb{R}$ provided that for each $x \in S$ there is an open interval of the form (a, b) such that $x \in (a, b) \subseteq S$. The empty set is $\varnothing$ also considered an open set.

Clearly, if $S = (a, b)$ is an open interval, then S is an open set because for each $x \in S$, we have that $x \in (a, b) \subseteq S = (a, b)$.

Definition 5 For a function $f : \mathbb{R} \to \mathbb{R}$ with domain D and $B \subseteq \mathbb{R}$, the **inverse image** of B, written $f^{-1}(B)$, is the set defined by

$$f^{-1}(B) = \{x \in D \mid f(x) \in B\}$$

Example 5 Let $f(x) = x^2$ and $B = (0,4)$, find $f^{-1}(B)$.

Solution By definition,

$$f^{-1}(B) = \{x \in \mathbb{R} \mid f(x) \in (0,4)\} = \{x \in \mathbb{R} \mid 0 < x^2 < 4\} = \{x \mid -2 < x < 0\} \cup \{x \mid 0 < x < 2\} \quad \blacksquare$$

Theorem 10 Let $f : \mathbb{R} \to \mathbb{R}$ with domain D, then f is continuous on D if and only if $f^{-1}(B)$ is open for every open set $B \subseteq \mathbb{R}$.

Proof First suppose that f is continuous on D.
Let B be an open subset of $\mathbb{R}$. Let $a \in f^{-1}(B)$, then f is continuous at $x = a$, since $a \in D$. Now $f(a) \in B$, which is an open set; thus, there is an open interval (c,d) with $f(a) \in (c,d) \subseteq B$. Choose $\epsilon > 0$ so that $(f(a) - \epsilon, f(a) + \epsilon) \subseteq (c,d)$. This is always possible since $c < f(a) < d$. Because f is continuous at a, there exist a $\delta > 0$ such that $|x - a| < \delta$ implies $|f(x) - f(a)| < \epsilon$, that is, $x \in (a - \delta, a + \delta)$ implies $f(x) \in (f(a) - \epsilon, f(a) + \epsilon)$.
Therefore $f[(a - \delta, a + \delta)] \subseteq (f(a) - \epsilon, f(a) + \epsilon) \subseteq B$. Consequently, the open interval $(a - \delta, a + \delta)$ is a subset of $f^{-1}(B)$ and $a \in (a - \delta, a + \delta) \subseteq f^{-1}(B)$. Since $a \in f^{-1}(B)$ was an arbitrary point, $f^{-1}(B)$ is an open set.

Next, suppose that $f^{-1}(B)$ is an open set for every open set $B \subseteq \mathbb{R}$.
Let $a \in D$ and choose any $\epsilon > 0$, then the open interval $B = (f(a) - \epsilon, f(a) + \epsilon)$ is an open subset and $f(a) \in B$. Thus, $a \in f^{-1}(B)$. Since by assumption $f^{-1}(B)$ is open, there is an open interval (c,d) with $a \in (c,d) \subseteq f^{-1}(B)$. Choose $\delta > 0$ such that $(a - \delta, a + \delta) \subseteq (c,d)$, then for $x \in (a - \delta, a + \delta)$, we have that $f(x) \in B$, since $(a - \delta, a + \delta) \subseteq (c,d) \subseteq f^{-1}(B)$. Thus, we have that $f[(a - \delta, a + \delta)] \subseteq B = (f(a) - \epsilon, f(a) + \epsilon)$. Consequently, we have shown that given any $\epsilon > 0$, there exists a $\delta > 0$ such that $f[(a - \delta, a + \delta)] \subseteq (f(a) - \epsilon, f(a) + \epsilon)$, that is, $|x - a| < \delta$ implies $|f(x) - f(a)| < \epsilon$. This shows that f is continuous at each $a \in D$. $\blacksquare$

The Intermediate Value Theorem and Connected Sets

Intuitively, we can describe a continuous function as a function whose graph does not have any breaks or gaps, that is, its graph can be drawn *without* lifting up the pencil when drawing its graph. This next theorem establishes the validity of this intuitive concept.

Theorem 11 (The Intermediate Value Theorem)
Let $[a,b]$ be a closed interval in $\mathbb{R}$ and let $f : \mathbb{R} \to \mathbb{R}$ be a continuous function on its domain with $[a,b]$ a subset of the domain of f. If y is any number between $f(a)$ and $f(b)$, then there exists a number x between a and b such that $f(x) = y$.

The proof of this theorem is based on the property of connectedness, which we now define.

Definition 6 Let A be a nonempty subset of $\mathbb{R}$. A is **disconnected**, if there are two nonempty open subsets U and V of $\mathbb{R}$ such that
(i) $A \subseteq U \cup V$
(ii) $A \cap U \neq \varnothing$ and $A \cap V \neq \varnothing$
(iii) $A \cap U \cap V = \varnothing$

We say that U and V disconnect A. A subset A is said to be **connected** if it is not disconnected.

Example 6 For any $a \in \mathbb{R}$, show that $A = \mathbb{R} - \{a\}$ is disconnected.

Proof Let $U = (-\infty, a)$ and $V = (a, +\infty)$. Then U and V are nonempty open subsets so that $A \subseteq U \cup V$. Now $A \cap U = (-\infty, a)$ and $A \cap V = (a, +\infty)$. Thus, $A \cap U \neq \varnothing$ and $A \cap V \neq \varnothing$. Moreover, $A \cap U \cap V = \varnothing$, since $U \cap V = \varnothing$. ■

To prove the Intermediate Value Theorem, we will need the following results.

Theorem 12 Let $f : \mathbb{R} \to \mathbb{R}$ be a continuous function on its domain D and let $A \subseteq D$. If A is connected, then its image $f(A)$ is connected.

Proof Assume that A is connected and $f : \mathbb{R} \to \mathbb{R}$ is continuous on its domain D with $A \subseteq D$.

Suppose that $f(A)$ is disconnected, then there are nonempty open sets U and V of $\mathbb{R}$ that disconnect $f(A)$. Because f is continuous, $f^{-1}(U)$ and $f^{-1}(V)$ are open sets. We will show that $f^{-1}(U)$ and $f^{-1}(V)$ disconnect A and obtain a contradiction to the hypothesis that A is connected.
Since U and V disconnect $f(A)$, we have the following:
(i) $f(A) \subseteq U \cup V$
(ii) $f(A) \cap U \neq \varnothing$ and $f(A) \cap V \neq \varnothing$
(iii) $f(A) \cap U \cap V = \varnothing$

For each $a \in A$, we have $f(a) \in f(A)$; thus, $A \subseteq f^{-1}(f(A)) \subseteq f^{-1}(U \cup V) = f^{-1}(U) \cup f^{-1}(V)$.
Because $f(A) \cap U \neq \varnothing$, we can show that $A \cap f^{-1}(U) \neq \varnothing$ as follows:
Let $y \in f(A) \cap U$, then $y \in f(A)$ and $y \in U$. Thus, there exist an $x \in A$ with $f(x) = y$. Since $y = f(x) \in U$, then $x \in f^{-1}(U)$ and consequently $x \in A \cap f^{-1}(U)$.
A similar argument shows that $A \cap f^{-1}(V) \neq \varnothing$, since $f(A) \cap V \neq \varnothing$.
Finally, $A \cap f^{-1}(U) \cap f^{-1}(V) = \varnothing$ because if $x \in A \cap f^{-1}(U) \cap f^{-1}(V)$, then we would have that $x \in A$ and $x \in f^{-1}(U)$ and $x \in f^{-1}(V)$ so that $f(x) \in f(A)$ and $f(x) \in U$ and $f(x) \in V$, which would make $f(x) \in f(A) \cap U \cap V$ and contradict $f(A) \cap U \cap V = \varnothing$.
Therefore, if (i),(ii), and (iii) are true, then $f^{-1}(U)$ and $f^{-1}(V)$ are nonempty open sets which disconnect A.
Thus, if we assume that $f(A)$ is disconnected, then we can show that A is disconnected. However, we are assuming that A is connected; thus, $f(A)$ must also be connected. ■

Recall that the intervals in $\mathbb{R}$ have the following form:
$(a,b), (a,b], [a,b), [a,b], (-\infty,a), (-\infty,a], (a,\infty), [a,\infty), (-\infty,\infty)$
An interval E in $\mathbb{R}$ can be characterized by the following property:

for any $a, b \in E$ if $a < x < b$, then $x \in E$

The connected sets of real numbers can be described completely by the following theorem.

Theorem 13 A subset E of $\mathbb{R}$ containing at least two points is connected iff E is an interval.

Proof Suppose that E is not an interval. Then, there exists $a,b \in E$ and $p \notin E$ such that $a < p < b$.
Let U be the open interval $(-\infty,p)$ and V be the open interval (p,∞) . Then $a \in U$ and $b \in V$ so that $E \cap U \neq \varnothing$ and $E \cap V \neq \varnothing$. Clearly, $E \subseteq U \cup V$ since the only real number not in $U \cup V$ is p and $p \notin E$. Also $E \cap U \cap V = \varnothing$, since $U \cap V = \varnothing$. Thus, U and V disconnect E .
The proof that " if E is an interval, then E is connected " is fairly complicated and is left as a sequence of exercises. ■

Theorem 13 establishes that the only connected subsets of $\mathbb{R}$ are intervals or sets of the form $\{a\}$ with $a \in \mathbb{R}$ (see Exercise 14). The Intermediate Value Theorem can now be proved as follows:

Proof of Theorem 11 Assume that f is continuous on $[a,b]$ and $f(a) < f(b)$ (the proof when $f(b) < f(a)$ is analogous).
Let y be any number with $f(a) < y < f(b)$ and suppose that $f(x) \neq y$, for every x with $a < x < b$.
Let $U = (-\infty,y)$ and $V = (y,\infty)$, then U and V are both open subsets of $\mathbb{R}$. Since $f(a) \in U$ and $f(b) \in V$, it follows that $f([a,b]) \cap U \neq \varnothing$ and $f([a,b]) \cap V \neq \varnothing$.
Because $f(x) \neq y$ for every x with $a < x < b$ and y is the only number not in $U \cup V$, we have that $f([a,b]) \subseteq U \cup V$. Moreover, $f([a,b]) \cap U \cap V = \varnothing$, since $U \cap V = \varnothing$.
Therefore, U and V disconnect $f([a,b])$. However, since the interval $[a,b]$ is connected and f is continuous on $[a,b]$, we have that $f([a,b])$ is connected by Theorem 12. Thus, if there exists a real number y with $f(a) < y < f(b)$ and $f(x) \neq y$, for every x with $a < x < b$, then we reach the contradiction that $f([a,b])$ is connected and disconnected. Consequently, for y any number with $f(a) < y < f(b)$, there must exist an x with $a < x < b$ such that $f(x) = y$. ■

Corollary 14 Let $[a,b]$ be a closed interval in $\mathbb{R}$ and let $f : \mathbb{R} \to \mathbb{R}$ be a continuous function on its domain with $[a,b]$ a subset of the domain of f . If $f(a)$ and $f(b)$ have opposite signs, then there exists an x with $a < x < b$ such that $f(x) = 0$.

The proof of Corollary 14 is left as an exercise. The next example shows the use of this corollary.

Example 7 The equation $x^5 + 2x - 5 = 0$ has a real solution between $x = 1$ and $x = 2$.

Proof Let $f(x) = x^5 + 2x - 5$, then $f(x)$ is continuous on $\mathbb{R}$, since $f(x)$ is a polynomial function. Observe that $f(1) = -2$ and $f(2) = 31$. Thus, by Corollary 14, there exist an s with $1 < s < 2$ such that $f(s) = s^5 + 2s - 5 = 0$. Hence, s is a solution. ■

Example 8 shows how to find the solution in Example 7 more accurately by "interval-halving".

Example 8 For $f(x) = x^5 + 2x - 5$, solve $f(x) = 0$ accurate to the nearest tenth.

Solution From Example 7, we know there is a solution s with $1 < s < 2$. Consider the midpoint $m_1 = 1.5$ between $x = 1$ and $x = 2$. Now $f(1.5) = 5.59375$ and $f(1) = -2$; thus, there must be a solution between $x = 1$ and $x = 1.5$. Continuing this process, consider the midpoint $m_2 = 1.25$ between $x = 1$ and $x = 1.5$. Now $f(1.25) = .55$ (approximately); thus, there must be a solution between $x = 1$ and $x = 1.25$. Continue with the midpoint $m_3 = 1.125$ between $x = 1$ and

$x = 1.25$. Now $f(1.125) = -.95$ (approximately). Thus, there must be a solution between $x = 1.125$ and $x = 1.25$, since $f(1.125)$ is negative and $f(1.25)$ is positive. Continue with the midpoint $m_4 = 1.1875$ between $x = 1.125$ and $x = 1.25$. Now $f(1.1875) = -.25$ (approximately). Thus, there must be a solution between $x = 1.1875$ and $x = 1.25$, since $f(1.1875)$ is negative and $f(1.25)$ is positive. Consequently, there is a solution given by $x = 1.2$ to the nearest tenth. This whole process could be carried out even more accurately by using a computer to implement the method of interval-halving. ■

Exercises

1. Use Definition 1 to prove that $\lim_{x\to 5}[3x - 4] = 11$.

2. Use Definition 1 to prove that $\lim_{x\to 2} x^2 = 4$.

3. Show the necessary steps to use the limit theorems to compute the following:
(a) $\lim_{x\to 2}[x^3 - 2x^2 + 5x]$
(b) $\lim_{x\to 5}[3x^2 - 2x][5x + 2]$
(c) $\lim_{x\to 1} \frac{2x^2+1}{3x-1}$

4. Use Mathematical Induction to prove Theorem 9.

5. Show that if S_1, S_2 are open sets then $S_1 \cup S_2$ is an open set.

6. Show that if (a,b) and (c,d) are open intervals with nonempty intersection, then $(a,b) \cap (c,d)$ is an open interval. (Hint: Consider various possible cases for a,b,c,d so that $(a,b) \cap (c,d) \neq \varnothing$).

7. Let $f(x) = x^3$ and $B = (0,27)$, find $f^{-1}(B)$.

8. Define $f : \mathbb{R} \to \mathbb{R}$ by $f(x) = 1$ for $x < 3$ and $f(x) = 2$ for $x \geq 3$.
Determine whether or not $\lim_{x\to 3} f(x)$ exists and verify your answer.

9. Let $p(x)$ and $q(x)$ be polynomial functions. Use the fact that $\lim_{x\to a} p(x) = p(a)$ and $\lim_{x\to a} q(x) = q(a)$ and appropriate limit theorems to prove that the rational function $\frac{p(x)}{q(x)}$ is continuous at $x = a$ provided that $q(a) \neq 0$.

10. Use Definition 1 to prove Theorem 2 part (b).

11. Assume that $\lim_{x\to a} f(x) = L_1$ and $\lim_{x\to a} f(x) = L_2$. Show that $L_1 = L_2$.
Hint: If $L_1 \neq L_2$ say $L_1 < L_2$, then choose $\epsilon > 0$ such that $L_1 + \epsilon < L_2 - \epsilon$ and hence $(L_1 - \epsilon, L_1 + \epsilon) \cap (L_2 - \epsilon, L_2 + \epsilon) = \varnothing$. Then use Definition 1 to show that a contradiction results for this choice of ϵ .

12. Prove Theorem 6.

13. Let a, b be any two points in $\mathbb{R}$, show that $\mathbb{R} - \{a, b\}$ is disconnected.

14. For any $a \in A$, show that $\{a\}$ can not be disconnected.

15. Give an example to show that the union of two connected sets is not necessarily connected.

16. Prove Corollary 14. Note that if $f(a)$ and $f(b)$ have opposite signs then either $f(a) < 0$ and $f(b) > 0$ or $f(a) > 0$ and $f(b) < 0$. Thus, 0 is always a number between $f(a)$ and $f(b)$ so that Theorem 11 can be used.

17. Show that $(-\infty, 0] \cup [6, 8]$ is disconnected.

18. Give a counterexample to the statement: If S and T are intervals, then $S \cup T$ is an interval.

19. Suppose that f is continuous on the interval $[a, b]$. Prove that if there exist $x_1, x_2 \in [a, b]$ with $f(x_1) \neq f(x_2)$, then $f([a, b])$ is an interval.

20. Let U be a nonempty subset of $\mathbb{R}$ that is bounded above. By the completeness property, U has a least upper bound m. Let $I = (c, d)$ be any open interval containing m, show that $I \cap U \neq \varnothing$.
Hint: Consider the subinterval (c, m), then show that if there is no element of U in (c, m), then c is an upper bound for U, which contradicts m as the least upper bound since $c < m$.

Solutions to Selected Exercises

Chapter 1 Exercises (page 5)

1(b). Assume that x is an odd integer and y is an even integer, then $x = 2m + 1$, for some integer m and $y = 2n$, for some integer n. Thus, $x + y = (2m + 1) + (2n) = 2m + 2n + 1 = 2(m + n) + 1$. Since $\mathbb{Z}$ is closed for addition, $m + n$ is an integer, say q. Then, $x + y = 2q + 1$ and therefore, $x + y$ is an odd integer.

1 (d). Assume that x is an odd integer, then $x = 2m + 1$, for some integer m. Thus, $x^2 = (2m + 1)(2m + 1) = 4m^2 + 4m + 1 = 2(2m^2 + 2m) + 1$. Since $\mathbb{Z}$ is closed for multiplication and addition, then $2m^2 + 2m$ is an integer, say q. Thus, $x^2 = 2q + 1$ and therefore x^2 is an odd integer.

1 (f). Assume that x is an odd integer and y is an even integer, then $x = 2m + 1$, for some integer m and $y = 2n$, for some integer n. Thus, $xy = (2m + 1)((2n) = 4mn + 2n = 2(2mn + n)$. Since $\mathbb{Z}$ is closed for addition and multiplication, then $2mn + n$ is an integer, say p. Thus, $xy = 2p$ and therefore xy is an even integer.

2 (b). Let $x = 12$ and $y = 4$ so that x and y are both even integers; however, $\frac{x}{y} = \frac{12}{4} = 3$ is odd.

Chapter 1 Exercises (page 9)

2. Assume that $ab = 0$. If $a \neq 0$, then a^{-1} exists. Now, $a^{-1}(ab) = a^{-1}(0)$, that is, $(a^{-1}a)b = 0$ so that $(1)b = 0$ and therefore $b = 0$. Similarly, if $b \neq 0$, then b^{-1} exists and it can be shown that $a = 0$. Thus, when $a \neq 0$ and $ab = 0$, then $b = 0$, and when $b \neq 0$ and $ab = 0$, then $a = 0$.

4. We use the result that $-x = (-1)x$ for $x \in \mathbb{R}$. Thus, $(-a)b = [(-1)a]b = a(-1)b = a(-b)$. Also, $(-a)b = [(-1)a]b = (-1)ab = -(ab)$.

6. For the stated proposition to make sense, it must be true that $b \neq 0$ and $d \neq 0$ so that b^{-1} and d^{-1} both exist. Assume that $ad = bc$, then $(ad)b^{-1} = (bc)b^{-1} = c(bb^{-1}) = c(1) = c$. Thus, $(ad)b^{-1} = c$ Next, multiply both sides of this equation by d^{-1} to obtain $[(ad)b^{-1}]d^{-1} = cd^{-1}$, that is $a(dd^{-1})b^{-1} = cd^{-1}$so that $a(1)b^{-1} = cd^{-1}$and therefore $ab^{-1} = cd^{-1}$, that is, $\frac{a}{b} = \frac{c}{d}$.

8. $(a + b)(c + d) = a(c + d) + b(c + d) = ac + ad + bc + bd = ac + bc + ad + bd$. Here, the first equality is obtained by applying the distributive property to the quantity $(c + d)$. The distributive and commutative properties are then used to simply further.

10. To prove the first part, we make use of Theorem 4 (iii) as follows: $(-b)^{-1} = [(-1)b]^{-1} = (-1)^{-1}(b^{-1}) = (-1)b^{-1} = -b^{-1}$. Here we also used the fact that $(-1)^{-1} = (-1)$, which is true because $(-1)(-1) = 1$ Using the definition of division and the above result $\frac{-a}{-b} = (-a)(-b)^{-1} = (-1)(a)(-1)(b^{-1}) = (-1)(-1)(ab^{-1}) = 1(ab^{-1}) = ab^{-1} = \frac{a}{b}$.

13. For the stated proposition to make sense, it must be true that $c \neq 0$ and $d \neq 0$ so that c^{-1} and

d^{-1} both exist. Consider $cd(c^{-1}d^{-1}) = (c)(c^{-1})(d)(d^{-1}) = (1)(1) = 1$. It can also be shown that $(c^{-1}d^{-1})(cd) = 1$. Thus,
$c^{-1}d^{-1}$ has the property of the multiplicative inverse of cd. Since the multiplicative inverse of any real number is unique, then $c^{-1}d^{-1}$ must be the multiplicative inverse of cd. Thus, $c^{-1}d^{-1})\backslash = (cd)^{-1}$.

16. By the definition of subtraction,
$\frac{a}{b} - \frac{c}{b} = \frac{a}{b} + [-\frac{c}{b}] = ab^{-1} + [-(c)b^{-1}] = ab^{-1} + (-c)(b^{-1} = [a + (-c)]b^{-1} = [a - c]b^{-1} = \frac{(a-c)}{b}$.

Chapter 1 Exercises (**page12**)

1 (b). We use a proof by contradiction: Assume that $a < 0$ and $a^{-1} \geq 0$. Case 1: $a^{-1} = 0$, then $a(a^{-1}) = 1 = 0$, which contradicts the fact the $1 \neq 0$. Case 2: $a^{-1} > 0$, then by Axiom 2, $a(a^{-1}) < 0$; however, $a(a^{-1}) = 1 > 0$ and thus a contradiction also results. Next, assume that $a < 0$, then by Definition 4, there is a $c > 0$ so that $a + c = 0$, By uniqiueness of additive inverses and the fact that $a + (-a) = 0$, it follows that $(-a) = c > 0$.

4. Assume that $a > b$ and $d > 0$. By Theorem 6, $d^{-1} > 0$. By Theorem 9, $a(d^{-1}) > b(d^{-1}$. By definition $ad^{-1} = \frac{a}{d}$ and $bd^{-1} = \frac{b}{d}$ and therefore $\frac{a}{d} > \frac{b}{d}$.

5. We use a proof by contradiction: Assume that $a > b$ and $d < 0$ and $ad \geq bd$. Case 1: $ad = bd$, then multiplying both sides of this equation by d^{-1} gives $(ad)d^{-1} = (bd)d^{-1}$, that is, $a(dd^{-1}) = b(dd^{-1})$ so that $a = b$ since $(dd^{-1}) = 1$. This contradicts that the assumption that $a > b$.
Now consider the case where $ad > bd$, that is, $ad - bd > 0$, also $d^{-1} < 0$, by Theorem 6. By Axiom 2, it follows that $(ad - bd)d^{-1} < 0$, that is $(ad)d^{-1} - (bd)d^{-1} < 0$, that is, $a(dd^{-1}) - b(dd^{-1}) < 0$ so that $a < b$, since $dd^{-1} = 1$. This result contradicts that the assumption that $a > b$. Therefore, $ad \geq bd$ can not be true, and $ad < bd$ must be true.

Chapter 2 Exercises (**page 23**)

2 (b)

P	Q	P ∨ Q	~Q	(P ∨ Q) ∧ ~Q
T	T	T	F	F
T	F	T	T	T
F	T	T	F	F
F	F	F	T	F

5 (b). The negation has the form $\sim p$ and $\sim q$. Thus, the negation is $\sim (x > 2)$ and $\sim (x < -5)$, that is, $(x \leq 2)$ and $(x \geq -5)$.

6 (b). If xy is not even, then x is not even or y is not even. Equivalently, If xy is odd, then x is odd or y is odd. Remember that the contapositive of P → Q has the form ~Q → ~P.

8.

P	Q	P → Q	~P	~P ∨ Q
T	T	T	F	T
T	F	F	F	F
F	T	T	T	T
F	F	T	T	T

Column 3 shows that the pattern of assigning truth values to P → Q is the same as the pattern for assigning values to ~P ∨ Q as shown in column 5.

11 (c). x is even only if x^2 is even

12. Use ~(P → Q) is equivalent to P ∧ ~Q to obtain the negation "x is a multiple of π and cos(x) $\neq 1$". When $x = 3\pi$, then cos(3π) $= -1 \neq 1$. Thus, $x = 3\pi$ is a counterexample to "If x is a multiple of π, then cos(x) =1".

14.

	P	Q	R	P	→	(Q ∧ R)	(P → Q)	∧	(P → R)
	T	T	T	T	T	T	T	T	T
	T	T	F	T	F	F	T	F	F
	T	F	T	T	F	F	F	F	T
	T	F	F	T	F	F	F	F	F
	F	T	T	F	T	T	T	T	T
	F	T	F	F	T	F	T	T	T
	F	F	T	F	T	F	T	T	T
	F	F	F	F	T	F	T	T	T
Step No.	1	1	1	2	4	3	2′	4′	3′

Column 4 shows that the pattern of assigning truth values to P → (Q ∧ R) is the same as the pattern of assigning truth values to (P → Q) ∧ (P → R) as shown in column 5.

16 (c). $x = -3$ if and only if $x^2 = 9$.

17. (Q ∨ R) → ~P is false only in the case that (Q ∨ R) is true and ~P is false. Since Q is false, then R is true in order for (Q ∨ R) to be true. Also, when ~P is false, then P is true.

Chapter 3 Exercises (page 33)

1. Assume that $a \mid b$ and $a \mid c$, then $am = b$ for some $m \in \mathbb{Z}$ and $an = c$ for some $n \in \mathbb{Z}$. Thus, $b - c = am - an = a(m - n)$. Since $\mathbb{Z}$ is closed for subtraction $m - n = q$ for some $q \in \mathbb{Z}$. Thus, $b - c = aq$ and therefore, $a \mid (b - c)$.

3. Assume that $a \mid b$ and c is a multiple of b. By definition, $am = b$ for some $m \in \mathbb{Z}$ and $c = bn$ for some $n \in \mathbb{Z}$. Thus, $c = (am)n = a(mn)$. Since $\mathbb{Z}$ is closed under multiplication, $mn = p$ for some

$p \in \mathbb{Z}$. Thus, $c = ap$ and therefore, $a \mid c$.

7. Let $a = 3$, $b = 6$, and $c = 6$. Then $a \mid c$ and $b \mid c$. However, $ab = 18$ is not a divisor of 6. Thus, $ab \nmid c$.

9. We prove the contrapositive: "If n is not odd, then n^2 is not odd.". Assume that n is not odd. Thus, n is even and $n = 2p$ for some $p \in \mathbb{Z}$. Then $n^2 = (2p)^2 = 4p^2 = 2(2p^2)$ so that $2 \mid n^2$. Consequently, n^2 is even and therefore n^2 is not odd is true.

15. Consider the consecutive integers n and $m = n + 1$. Then, $m^2 + n^2 - 1 = (n+1)^2 + n^2 - 1 = n^2 + 2n + 1 + n^2 - 1 = 2n^2 + 2n = 2n(n+1)$. Case 1: n is odd so that $n = 2p + 1$ for some $p \in \mathbb{Z}$. In this case, $2n(n+1) = 2n(2p+2) = 4n(p+1)$ so that $4 \mid (m^2 + n^2 - 1)$. Case 2: n is even and $n = 2q$ for some $q \in \mathbb{Z}$. In Case 2, $2n(n+1) = 2(2q)(2q+1) = 4q(2q+1)$. Thus, $4 \mid (m^2 + n^2 - 1)$.

23. Assume $a \approxeq 5 (\mathrm{mod}\, 6)$ then $a - 5 = 6n$ for some $n \in \mathbb{Z}$. Thus, $a = 6n + 5$ and $a^2 = 36n^2 + 60n + 25$. Then, $a^2 - 1 = 36n^2 + 60n + 24 = 6(6n^2 + 10n + 4)$. Consequently, $6 \mid (a^2 - 1)$ so that $a^2 \approxeq 1 (\mathrm{mod}\, 6)$.

28. Let x be a positive integer then $x \in \mathbb{N}$. If $x = 1$, then $2x = 2$ and $x^2 = 1$. Thus, $2x < x^2 < 3x$ is false. If $x = 2$, then $x^2 = 4 = 2x$. Thus, $2x < x^2 < 3x$ is false. If $3 \leq x$, then multiplying both sides of this inequality by x gives $3x \leq x^2$, since $x > 0$. Thus, $2x < x^2 < 3x$ is false. Since $\mathbb{N} = \{1,2\} \cup \{x \mid 3 \leq x, x \text{ is an integer}\}$, we have shown that for $x \in \mathbb{N}$, the inequality $2x < x^2 < 3x$ is never true.

Chapter 4 Exercises (page 37)

12 (a). $\{x \in \mathbb{N} \mid x \text{ is even}\}$

13. The 8 subsets are $\varnothing$, $\{a\}$, $\{b\}$, $\{c\}$, $\{a,b\}$, $\{a,c\}$, $\{b,c\}$, $\{a,b,c\}$.

16. Assume that $A \subseteq B$ and $B \subseteq C$. Let $x \in A$, then $x \in B$, because $A \subseteq B$ means that every element of A is also an element of B. Now, when $x \in B$ then $x \in C$ follows, since $B \subseteq C$. Thus, we have shown for any element x in A it is true that x is also in C, and consequently $A \subseteq C$.

Chapter 4 Exercises (page 41)

10 (a). Let $A = \{1,2\}$, $B = \{1,2,3\}$, and $C = \{1,2,3,4\}$. Then $A \cap B = \{1,2\} = A \cap C$. However, $B \neq C$.

12. Let $x \in (A \cap B)'$ then $x \notin (A \cap B)$. By definition, $A \cap B = \{y \mid y \in A \text{ and } y \in B\}$; thus, the statement $x \in A$ and $x \in B$ is false. By DeMorgan's Law, the negation of a p and q statement is equivalent to $\sim p$ or $\sim q$. Thus, the negation of $x \in A$ and $x \in B$ is $x \notin A$ or $x \notin B$, that is, $x \in A'$ or $x \in B'$, that is, $x \in A' \cup B'$. We have shown that the statement $x \in (A \cap B)'$ is equivalent to the statement $x \in A' \cup B'$, consequently, $(A \cap B)' = A' \cup B'$.

17. It is easier to prove the contrapositive "If $A \not\subseteq B'$, then $A \cap B \neq \varnothing$". Assume that $A \not\subseteq B'$, then $\exists$ $x \in A$ such that $x \notin B'$. Since $B \cup B' = U$, it follows that $x \in B$. Thus, $x \in A$ and $x \in B$, so that $x \in A \cap B$. Therefore, $A \cap B \neq \varnothing$.

20. The proof will be a proof by contradiction. Assume the hypothesis that $A \not\subseteq B$ and $C \subseteq B$ are true and the conclusion $A \not\subseteq C$ is false. If $A \not\subseteq C$ is false, then $A \subseteq C$ is true. Since $C \subseteq B$, then $A \subseteq B$ must be true also. Thus, for every $x \in A$, it follows that $x \in B$. Therefore, $A \not\subseteq B$ is false. This contradicts that $A \not\subseteq B$ was assumed to be true.

Chapter 5 Exercises (**page 50**)

1 (a). $(\forall\ x,y \in S)\ (x - y \in S)$

1 (c). $(\forall\ n \in \mathbb{Z})$ ($n + (n+1) + (n+2)$ is divisible by 3)

1 (f). $(\forall\ x \in \mathbb{R})$ ($\exists\ y \in \mathbb{R}$ such that $x + y = 0$)

2 (a). $(\forall\ a \in \mathbb{N})$ ($b \mid a$ and $c \mid a$ and $(bc \nmid a)$) Remember that the negation of $p \rightarrow q$ is p and $\sim q$.

2 (d). ($\exists\ m \in \mathbb{Z}$) ($m^2 + m$ is even)

3. Let $a = 2$, $b = 3$, and $c = 5$, then for $x = 1$ and $y = 1$, $ax + by = c$ is true. Let $a = 2$, $b = 4$, and $c = 7$, then $\forall\ x, y \in \mathbb{Z}$, $ax + by$ is even. thus, $ax + by = c$ is false for $\forall\ x, y \in \mathbb{Z}$, since c is odd.

Chapter 5 Exercises (**page 57**)

2. Any three consecutive integers can be written as n, $n+1$, $n+2$ for some $n \in \mathbb{Z}$. Thus, the sum S of these is given by $S = n + (n+1) + (n+2) = 3n + 3 = 3(n+1)$. Consequently, $3 \mid S$.

5. Let $f(x) = 4x + 2$ for $x \in \mathbb{R}$. If $f(x) = f(y)$ for some $x,\ y \in \mathbb{R}$ then $4x + 2 = 4y + 2$. Thus, $4x = 4y$ and therefore, $x = y$. This shows that f is one-to-one. Next, let $y \in \mathbb{R}$. To find x such that $f(x) = y$, solve the equation $y = 4x + 2$ for x as follows: $y - 2 = 4x$ and therefore $\frac{y-2}{4} = x$. Since $\mathbb{R}$ is closed under subtraction and division by a nonzero number, we have $x \in \mathbb{R}$ and $f(x) = y$. Since y is an arbitrary real number, f is onto.

7. Let S be the interval $[a,\ m]$ for a, m positive real numbers with $a < m$, and let T be the interval $[a, \infty)$, then T is not bounded above; however, S is bounded above by m and $S \subset T$. This counterexample disproves the given statement.

9. Assume $x, y \in \mathbb{R}$ and $x \neq y$. Case 1: $x < y$, then $f(x) < f(y)$ since f is strictly increasing. Case 2: $y < x$, then $f(y) < f(x)$. Case 1 and Case 2 show that whenever $x \neq y$, then $f(x) \neq f(y)$. Therefore, f is one-to-one.

13. Let $n \in \mathbb{Z}$ such that $n^4 = (n^2)^2$ is even. By the result that "If p^2 is even then p is even", we have that n^2 is even. Now n^2 is even implies that n is even. Since $3n$ is a multiple of an even number, $3n$

is even. Thus, $3n = 2q$ for some $q \in \mathbb{Z}$. Therefore, $3n + 1 = 2q + 1$ is odd.

19. Hint: Find a counterexample, by constructing sets A, B, C, and D so that (1) $A \subset B$, $C \subset D$ and $A \cap C = \varnothing$, and (2) $B - A$ and $D - C$ have at least one element in common.
24. If $a = 0$, then $f(x) = x^2 + b$. Let $x = -2$ and $y = 2$, then $f(-2) = 4 + b$ and $f(2) = 4 + b$ so that $f(x) = f(y)$. Therefore, $f(x)$ is not one-to-one. If $a \neq 0$, then $f(x) = x^2 + ax + b$ is a parabola with vertex at $x = \frac{-a}{2}$. The vertical line $x = \frac{-a}{2}$ is the axis of symmetry for this parabola. Thus, any two values of x, which have the same distance from the axis of symmetry and are on different sides of the axis of symmetry, will give the same $f(x)$ value. Thus, $f(x)$ is not one-to-one.

25 (b). $f \circ g(x) = 7(3x^2 + 1) + 3 = 21x^2 + 10$

Chapter 6 Exercises (page 67)

1 (c). For $n = 1$, $\frac{3(1)(1+1)}{2} = 3$ so that P(1) is true.
Assume P(k) is true, that is, $3 + 6 + 9 + \ldots + 3k = \frac{3k(k+1)}{2}$. Consider $S = 3 + 6 + 9 + \ldots + 3k$ so that $S + 3(k+1) = \frac{3k(k+1)}{2} + 3(k+1) = \frac{3k(k+1)}{2} + \frac{3(k+1)2}{2} = \frac{(k+1)[3k+3(2)]}{2} = \frac{(k+1)([3(k+2)]}{2} = \frac{3(k+1)(k+2)}{2}$
Thus, $S + 3(k+1) = \frac{3(k+1)(k+2)}{2}$ and P(k+1) is true. Therefore, P(1) is true and P(k) implies P(k+1) is true. Hence, P(n) is true for $n \in \mathbb{N}$.

1 (g). For $n = 1$, $5^1 - 2^1 = 3$ is a multiple of 3 so that P(1) is true. Assume P(k) is true, that is, $5^k - 2^k$ is a multiple of 3. Now,
$(5^k - 2^k)(5) = 5^k(5) - 2^k(5) = 5^{k+1} - 2^k(2+3) = 5^{k+1} - 2^k(2) - 2^k(3) = 5^{k+1} - 2^{k+1} - 2^k(3)$. Thus, $5^{k+1} - 2^{k+1} = 2^k(3) + (5^k - 2^k)(5)$, which is a multiple of 3, since $(5^k - 2^k)(5)$ is a multiple of 3. This demonstrates that when P(k) is true, then P(k+1) can be shown to be true. Hence, P(n) is true for $n \in \mathbb{N}$.

3 (c). For $n = 9$, $4^9 = 262144$ and $9! = 362880$. Thus, P(9) is true. Assume for $k > 9$ that P(k) is true, that is, $4^k \leq k!$. Now, $4^k(4) \leq (4)k! \leq (k+1)(k!)$, since $0 \leq 4$ and $4 \leq 10 < (k+1)$. Therefore, $4^{k+1} \leq (k+1)!$ since $4^{k+1} = 4^k(4)$ and $(k+1)! = (k+1)(k!)$. Thus, P(k) implies P(k+1) is true. Since P(1) is true, then Pn) is true for $9 \leq n$.

4 (b). If $0 < a < b$, then $0 < a^1 < b^1$, by the rules of exponents. Thus, P(1) is true. Assume that P(k) is true, that is, $0 < a^k < b^k$ for $0 < a < b$. Multiplying the inequality $0 < a^k < b^k$ by a gives $0 < a(a^k) < a(b^k)$ since $0 < a$. Now $a(b^k) < b(b^k)$ since $0 < a < b$. It follows that $0 < a(a^k) < b(b^k)$, that is, $0 < a^{k+1} < b^{k+1}$. This shows that P(k) implies P(k +1) is true. Since P(1) is true, then P(n) is true for $n \in \mathbb{N}$.

10. For $n = 7$, $n = 2 + 5$ so that P(7) is true. Assume for $7 < k$, that P(k) is true, that is k is a sum of 2's and 5's.
Case 1: k uses only 2's in its sum. Then $k = 2y + 2$, for some $y \in \mathbb{N}$ with $3 \leq y$, since $8 \leq k$. In this case, $k + 1 = 2y + 2 + 1 = 2(y-1) + 2 + 2 + 1 = 2(y-1) + 5$ so that $(k+1)$ is a sum of 2's and 5's.
Case 2: k uses only 5's in its sum; thus, $k = 5x$ for some $x \in \mathbb{N}$, with $2 \leq x$, since $10 \leq k$. In this case, $k + 1 = 5x + 1 = 5(x-1) + 5 + 1 = 5(x-1) + 6 = 5(x-1) + 3(2)$ so that $k + 1$ is a sum of 2's and 5's.
Case 3: k has at least one 5 and at least one 2 in its sum. Then $k = 5x + 5y$ for $x, y \in \mathbb{N}$. In this case,

$k+1 = 5x+5y+1 = 5(x-1)+2(y-1)+5+2+1 = 5(x-1)+2(y-1)+8$ so that $k+1$ is a sum of 2's and 5's, since $8 = 4(2)$.
Therefore, P(1) is true and P(k) implies P(k+1) is true. Hence, P(n) is true for $7 \le n$.

Chapter 7 Exercises (page 81)

3 (a). Domain of $R = \{x \mid x^2 \le 25\}$. Observe that if $x^2 > 25$, then there is no real number y such that $x^2 + y^2 \le 25$.

3 (b). Image of $R = \{y \mid y^2 \le 25\}$ because if $y^2 \le 25$ then $\exists\, x \in \mathbb{R}$ such that $x^2 + y^2 = 25$, that is, $(x,y) \in R$.

3 (c). R is not a function since $(3,4)$ and $(3,-4)$ are both elements of R.

5 (a). Let $a \in \mathrm{Im}(R^{-1})$ then there is a $b \in B$ such that $(b,a) \in R^{-1}$, that is, $(a,b) \in R$, by definition of R^{-1}. Thus, $a \in Dom(R)$ so that $\mathrm{Im}(R^{-1}) \subseteq Dom(R)$. Next, let $a_1 \in Dom(R)$, then there is a $b_1 \in B$ such that $(a_1,b_1) \in R$. Thus, $(b_1,a_1) \in R^{-1}$, that is, $a_1 \in \mathrm{Im}(R^{-1})$ so that $Dom(R) \subseteq \mathrm{Im}(R^{-1})$. Therefore, $\mathrm{Im}(R^{-1}) = Dom(R)$.

8 (a). ~ is reflexive since for $x \in \mathbb{Z}$, $x \cdot x \ge 0$ so that $x \sim x$.

8 (b). Let $x \sim y$ then $x \cdot y \ge 0$, that is $y \cdot x \ge 0$ so that $y \sim x$. Therefore, ~ is symmetric.

8 (c). Let $x \sim y$ and $y \sim z$, then $x \cdot y \ge 0$ and $y \cdot z \ge 0$. If $x \ge 0$, then $y \ge 0$ and consequently, $z \ge 0$ so that $x \cdot z \ge 0$. Thus, in this case, $x \sim z$.
If $x < 0$, then $y < 0$ since $x \cdot y \ge 0$. Consequently, $z < 0$, since $y \cdot z \ge 0$. Thus, in this case, $x \cdot z \ge 0$ so that $x \sim z$. Therefore, ~ is transitive.

!0 (a). For $A \ne \varnothing$, then $A \cap A \ne \varnothing$. Thus, ~ is not reflexive. If $A \sim B$, then $A \cap B = \varnothing$, that is, $B \cap A = \varnothing$ so that $B \sim A$. Thus, ~ is symmetric.
Consider $A = \{1,2\}$, $B = \{3,4\}$, and $C = \{1,5\}$. Then, $A \cap B = \varnothing$, so that $A \sim B$. Also, $B \cap C = \varnothing$ so that $B \sim C$. However, $A \cap C \ne \varnothing$ so that $A \sim C$ is not true and therefore ~ is not transitive.

10 (e). Observe that $l_1 \sim l_2$ means that the line l_1 is not parallel to line l_2. Since any line intercects itself in infinitely many points, then ~ is not reflexive.
If $l_1 \sim l_2$ then $l_1$intersects l_2 in a point so that l_2 intersects l_1 in the same point. Thus, $l_2 \sim l_1$ and ~ is symmetric.
Let l_1 and l_3 be any two lines that are parallel so that $l_1 \sim l_3$ is false. Now let l_2 be any line with $l_2 \sim l_1$, then l_2 is not parallel to l_1 and therefore l_2 is also not parallel to l_3 so that l_2 and l_3 must intersect in a point. In this case, $l_1 \sim l_2$ and $l_2 \sim l_3$, yet $l_1 \sim l_3$ is false. Thus, ~ is not transitive.

12 (a). Since $3 \mid (x+2x) = 3x$, then $x \sim x$ and ~ is reflexive. Let $x \sim y$ so that $3 \mid (x+2y)$, then $\exists$ $m \in \mathbb{Z}$ such that $3m = x+2y$. Thus, $6m = 2x+4y$, that is, $6m - 3y = 2x+y$ so that $3(2m-y) = y+2x$. Thus, $3 \mid y+2x$ and $y \sim x$ and therefore ~ is symmetric. Similar reasoning can be used to show that ~ is transitive.

12 (b). Observe that $0 \sim 3$ and $0 \sim 6$. Thus, the equivalence class $[0] = \{0,3,6\}$. Also, $1 \sim 4$ and $1 \sim 7$. Thus, the equivalence class $[1] = \{1,4,7\}$ Finally, $2 \sim 5$ so that $[2] = \{2,5\}$

15 (b). Let $(a,b) \in \mathbb{R} \, x \, \mathbb{R}$, then the equation $\frac{y-b}{x-a} = 2$, that is, $y - b = 2(x-a)$ gives a line with slope 2 that contains the point (a,b). Observe that every line with slope 2 can be written in slope-intercept form as $y = 2x + b$, for some $b \in \mathbb{R}$. If l_1 is the line $y = 2x + b_1$ and l_2 is the line $y = 2x + b_2$, then when $b_1 \neq b_2$, line l_1 is parallel to line l_2 so that these lines have no point in common. If $b_1 = b_2$, then the lines coincide and $l_1 = l_2$. Therefore, $\mathbb{C}$ is a partition of $\mathbb{R} \, x \, \mathbb{R}$.

17 (a). The equivlence classes are $[a] = \{a,b\}$, $[c] = \{c,d\}$, and $[e] = \{e\}$. Thus, the partition is $\{\{a,b\}\ ,\ \{c,d\}\ ,\ \{e\}\ \}$.

18 (a). Let $u = \{a,b,c\}$ and $V = \{d,e,f\}$, then the equivalence relation R that corresponds to this partition is defined by $x \sim y$ if $x \in U$ and $y \in U$ or if $x \in V$ and $y \in V$. Let $S = \{(x,y) \mid x \in U$ and $y \in U\}$ and let $T = \{(x,y) \mid x \in V$ and $y \in V\}$, then
$S = \{(a,a),(b,b),(c,c),(a,b),(b,a),(b,c),(c,b),(a,c),(c,a)\}$ and
$T = \{(d,d),(e,e),(f,f),(d,e),(e,d),(d,f),(f,d),(e,f),(f,e)\}$. Therefore, $R = S \cup T$.

Chapter 8 Exercises (page 93)

3 (a). $\lim_{x\to 2} x^3 - \lim_{x\to 2} 2x^2 + \lim_{x\to 2} 5x = (\lim_{x\to 2} x)^3 - 2(\lim_{x\to 2} x)^2 + 5\lim_{x\to 2} x = (2)^3 - 2(2)^2 + 5(2) = 8 - 8 + 10 = 10$

5. Assume that S_1, S_2 are open sets, let $x \in S_1 \cup S_2$, then $x \in S_1$ or $x \in S_2$. If $x \in S_1$, then there is an open interval (a,b) such that $x \in (a,b) \subseteq S_1$ since S_1 is open. Thus, $x \in (a,b) \subseteq S_1 \cup S_2$. If $x \in S_2$, then there is an open interval (c,d) such that $x \in (c,d) \subseteq S_2$ since S_2 is open. Thus, $x \in (c,d) \subseteq S_1 \cup S_2$. We have shown that for each $x \in S_1 \cup S_2$, there is an open interval containing x such that this open interval is a subset of $S_1 \cup S_2$. Hence, $S_1 \cup S_2$ is an open set.

8. Define $f : \mathbb{R} \to \mathbb{R}$ such that $f(x) = 1$, for $x < 3$ and $f(x) = 2$, for $3 \leq x$. Then the $\lim_{x\to 3} f(x)$ does not exist. Given $0 < \epsilon < 0.5$ and any $\delta > 0$, then the interval $S = (3-\delta, 3)$ is mapped by f into the inteval $(1-\epsilon, 1+\epsilon)$ whereas the interval $T = (3, 3+\delta)$ is mapped by f into the inerval $(2-e, 2+\epsilon)$. Because $1 + \epsilon < 2 - \epsilon$, it follows that $f(S \cup T) \not\subseteq (1-\epsilon, 1+\epsilon)$ so that $\lim_{x\to 3} f(x) \neq 1$ and $f(S \cup T) \not\subseteq (2-\epsilon, 2+\epsilon)$ so that $\lim_{x\to 3} f(x) \neq 2$. Observe that if $L \neq 1$ or $L \neq 2$, then choose $\epsilon > 0$ so that $1 \notin (L-\epsilon, L+\epsilon)$ and $2 \notin (L-\epsilon, L+\epsilon)$, then for any $\delta > 0$ with S and T as above, $f(S \cup T) \not\subseteq (L-e, L+\epsilon)$ so that L can not be $\lim_{x\to 3} f(x)$.

13. We can assume that $a < b$, then $A = \mathbb{R} - \{a,b\} = (-\infty, a) \cup (a,b) \cup (b,\infty)$. Let $U = (-\infty, a)$ and $V = (a,b) \cup (b,\infty)$, then U and V are open sets such that $A \subseteq U \cup V$ and $A \cap U \neq \varnothing$ and $A \cap V \neq \varnothing$. Since $U \cap V = \varnothing$, then $A \cap U \cap V = \varnothing$. Therefore, A is disconnected.

18. Consider the intervals $S = (1,3)$ and $T = (4,6)$. Let $a = 2$ and $b = 5$, then $a \in S \cup T$ and $b \in S \cup T$; however, for $x = 3.5$, then $a < x < b$, yet $x \notin S \cup T$. Thus, $S \cup T$ is not an interval.

APPENDIX A – Division Algorithm

For natural numbers a and b, to divide b by a means finding a quotient q and remainder r. For example, if $a = 5$ and $b = 23$, we have that $23 = 5 \cdot 4 + 3$ so that $q = 4$ and $r = 3$. We can also write $23 = 5 \cdot 3 + 8$ so that $q = 3$ and $r = 8$; however, $r > a$ in this case. Observe that when $r = 0$, then a is a divisor of b. In general, for $a, b \in \mathbb{N}$, with $a < b$, it is always possible to write $b = aq + r$, for some $q \in \mathbb{N}$ and $0 \leq r < a$. The number q is called the quotient and the number r is called the remainder. This result is an important theorem called the *Division Algorithm.*

Theorem I (**Division Algorithm**) For $a, b \in \mathbb{N}$ with $a < b$, there exist unique integers q and r such that $b = aq + r$ and $0 \leq r < a$.

When we perform the operation of dividing b by a, our goal is to make the remainder $r = b - aq$ as small as possible. With this in mind, we proceed with the proof of Theorem I as follows:

Proof of Theorem I

Consider the set $T = \{b - ax : x \in \mathbb{Z} \text{ and } 0 \leq b - ax\}$.

Now $b = b - a \cdot 0$ and $b > 0$; thus, $b \in T$ so that T is a nonempty subset of the natural numbers $\mathbb{N}$. The Well-Ordering Axiom for $\mathbb{N}$ states that every nonempty subset of $\mathbb{N}$ has a smallest element. Therefore, T has a smallest element, which is denoted by r. Since $r \in T$, there is some integer q such that $r = b - aq$ and $0 \leq r$.

Next, we show that $0 \leq r < a$. Suppose instead that $a \leq r$. Let $s = r - a$ so that $s \geq 0$. Because $a > 0$, it follows that $s < r$. Observe that

$$s = r - a = (b - aq) - a = b - (aq + a) = b - a(q + 1)$$

Since $s \geq 0$, the above expression shows that $s \in T$. When we assume that $a \leq r$ then it follows that $s < r$ and $s \in T$ and this contradicts the fact that r is the smallest element in T. Consequently, $a \leq r$ is not true and $r < a$ must be true.

To show that q and r are unique means showing that q and r are the only integers for which $b = aq + r$ and $0 \leq r < a$.

Suppose that q_2 and r_2 are integers such that $b = aq_2 + r_2$, where $0 \leq r_2 < a$. We may assume that $r_2 \geq r$ so that $r_2 - r \geq 0$. Since $b = aq + r$ and $b = aq_2 + r_2$, it follows that $aq + r = aq_2 + r_2$. Thus, $aq - aq_2 = r_2 - r$, that is, $a(q - q_2) = r_2 - r$ so that $a \mid (r_2 - r)$. Observe that $0 \leq (r_2 - r) < a$ since $0 \leq r_2 < a$ and $0 \leq r < a$ and $r_2 \geq r$. Consequently, $r_2 - r = 0$, that is, $r_2 = r$, by the result that states if $a \mid m$, then $m = 0$ or $m \geq a$. Furthermore, because $r_2 - r = 0$, then $a(q - q_2) = r_2 - r = 0$. Therefore, $q - q_2 = 0$, since $a \neq 0$. Thus, $q = q_2$ is also true and consequently, q and r are unique.

In Theorem I, a and b are positive integers. We can prove the following more general result.

Theorem II (**General Division Algorithm**) For integers a and b, with $a \neq 0$, there exist unique integers q and r such that $b = aq + r$ and $0 \leq r < ABS(a)$.

We now explore some aplications of the Division Algorithm:

Consider $a = 2$ and b any integer, then by the Division Algorithm, $b = aq_1 + r_1$ or $b = aq_2 + r_2$, where $r_1 = 0$ or $r_2 = 1$. When $r_1 = 0$, then $b = 2q_1$ so that b is an even integer. When $r_2 = 1$, then $b = 2q_2 + 1$ so that b is an odd integer. This result verifies that given any integer, it is either an even integer or it is an odd integer.

Next, consider $a = 3$ and any integer b. By the Division Algorithm, $b = 3q + r$ for some integers q and r with $0 \leq r < 3$, so that r must be $0, 1$ or 2.

For $r = 0$, if $b = 3q + 0$ for some q, then b is a multiple of 3. Also, observe that $b \approxeq 0 \bmod 3$ and b is in the equivalence class $[0]$. Recall that $[0]$ is the set of all integers, which are equivalent to 0 for the equivalence relation $x \sim y$ if and only if $x \approxeq y \bmod 3$.

For $r = 1$, if $b = 3q + 1$ for some q, then $b - 1 = 3q$. This means that $b \approxeq 1 \bmod 3$ and b is in the equivalence class $[1]$. Recall that $[1]$ is the set of all integers, which are equivalent to 1 for the equivalence relation $x \sim y$ if and only if $x \approxeq y \bmod 3$.

For $r = 2$, if $b = 3q + 2$ for some q, then $b - 2 = 3q$. This means that $b \approxeq 2 \bmod 3$ and b is in the equivalence class $[2]$. Recall that $[2]$ is the set of all integers, which are equivalent to 2 for the equivalence relation $x \sim y$ if and only if $x \approxeq y \bmod 3$.

Consequently, $b \in [0] \cup [1] \cup [2]$. Since the equivalence classes $[0]$, $[1]$, and $[2]$ are pairwise disjoint, these classes form a partition of $\mathbb{Z}$.

Next, consider $a = 4$ and any integer b. By the Division Algorithm, $b = 4q + r$ for some integers q and r with $0 \leq r < 4$, so that r must be $0, 1, 2$ or 3.

For $r = 0$ and $b = 4q + 0$, then b is in the equivalence class $[0]$, for the equivalence relation $x \sim y$ if and only if $x \approxeq y \bmod 4$.

For $r = 1$ and $b = 4q + 1$, then b is in the equivalence class $[1]$, for the equivalence relation $x \sim y$ if and only if $x \approxeq y \bmod 4$.

For $r = 2$ and $b = 4q + 2$, tben b is in the equivalence class $[2]$, for the equivalence relation $x \sim y$ if and only if $x \approxeq y \bmod 4$.

For $r = 3$ and $b = 4q + 3$, then b is in the equivalence class $[3]$, for the equivalence relation $x \sim y$ if and only if $x \approxeq y \bmod 4$.

Consequently, $b \in [0] \cup [1] \cup [2] \cup [3]$. Since the equivalence classes $[0]$, $[1]$, $[2]$ and $[3]$ are pairwise disjoint, these classes form a partition of $\mathbb{Z}$.

We generalize the previous examples as follows:

For any integer $n \geq 2$, by the Division Algorithm, if $b \in \mathbb{Z}$, then $b = nq + r$ for some integers r and q with $0 \leq r < n$. Thus, $r \in \{0, 1, 2, \ldots, n - 1\}$ and b is in one of the equivalence classes $[0]$, $[1]$, …, $[n - 1]$ for the equivalence relation $x \sim y$ if and only if $x \approxeq y \bmod n$. Thus, $\mathbb{Z} = [0] \cup [1] \cup \ldots \cup [n - 1]$. Since the equivalence classes $[0]$, $[1]$, …, $[n - 1]$ are pairwise disjoint, these classes form a partition of $\mathbb{Z}$.

For each $r = 0, 1, \ldots, n - 1$, the equivalence class $[r] = \{nq + r : q \in \mathbb{Z}\}$, that is $[r]$ consists of all those integers having a remainder of r when divided by n. These equivalence classes are called the residue classes modulo n.

We can now discuss an important method for finding the greatest common divisor of two integers. This method, which is called the Euclidean Algorithm, makes use of repeated applications of the Division Algorithm and the following theorem.

Theorem III Let $a, b \in \mathbb{N}$, with $0 < a \leq b$. If $b = aq + r$ for some integers q and r, then $\gcd(a, b) = \gcd(r, a)$.

Proof: Let $d = \gcd(a, b)$ and $h = \gcd(a, r)$. Because $b = aq + r = aq + r \cdot 1$, then b is a linear combination of a and r. Since $h = \gcd(a, r)$, by definition $h \mid a$ and $h \mid r$. Thus, $h \mid (aq + r \cdot 1)$ so that $h \mid b$. Hence, h is a common divisor of a and b. Therefore, $h \leq d$, because d is the greatest common divisor of a and b.

Using $b = aq + r$, we have $r = b - aq = b \cdot 1 + a(-q)$. Thus, r is a linear combination of b and a. Because $d = \gcd(b, a)$, then $d \mid b$ and $d \mid a$. Thus, $d \mid [b \cdot 1 + a(-q)]$, that is, $d \mid r$. Therefore, d is a common divisor of r and a. Since $h = \gcd(r, a)$, it follows that $d \leq h$. We have shown that $h \leq d$ and $d \leq h$, consequently, $d = h$, that is, $\gcd(a, b) = \gcd(r, a)$.

The Euclidean Algorithm can now be described. Let a and b be integers with $0 < a \leq b$. If $a \mid b$, then $\gcd(a, b) = a$. If $a \nmid b$, we apply the Division Algorithm and Theorem III repeatedly until a remainder of 0 is obtained. The last nonzero remainder is then $\gcd(a, b)$. The following example illustrates this mehod.

Example 1: Use the Euclidean Algorithm to find $\gcd(1155, 252)$.
Solution:
Dividing 1155 by 252 gives $1155 = 252(4) + 147$.
By Theorem III, $\gcd(1155, 252) = \gcd(252, 147)$
Dividing 252 by 147 gives $252 = 147 + 105$.
Thus, $\gcd(252, 147) = \gcd(147, 105)$.
Dividing 147 by 105 gives $147 = 105 + 42$.
Thus, $\gcd(147, 105) = \gcd(105, 42)$.
Dividing 105 by 42 gives $105 = 42(2) + 21$.
Thus, $\gcd(105, 42) = \gcd(42, 21)$.
Dividing 42 by 21 gives $42 = 21(2) + 0$ so that $\gcd(42, 21) = 21$.
Therefore, $\gcd(1155, 252) = \gcd(252, 147) = \gcd(147, 105) = \gcd(105, 42) = \gcd(42, 21) = 21$.

Using the prime factorization method, we have $1155 = 3 \cdot 5 \cdot 7 \cdot 11$ and $252 = 2 \cdot 2 \cdot 3 \cdot 3 \cdot 7$. Thus, the $\gcd(1155, 252) = 3 \cdot 7 = 21$.

INDEX

Made in the USA
San Bernardino, CA
17 April 2020